CONTRA TECHNOLOGIAM

The Crisis of Value in a Technological Age

Theodore John Rivers

UNIVERSITY
PRESS OF
AMERICA

Lanham • New York • London

Copyright © 1993 by
University Press of America®, Inc.
4720 Boston Way
Lanham, Maryland 20706

3 Henrietta Street
London WC2E 8LU England

All rights reserved
Printed in the United States of America
British Cataloging in Publication Information Available

Library of Congress Cataloging-in-Publication Data

Rivers, Theodore John.
Contra technologiam : the crisis of value in a technological age / by
Theodore John Rivers.
p.   cm. — (Toward a philosophy of culture ; v. 2)
Includes bibliographical references.
1. Technology and civilization.   2. Technology—Social aspects.
I. Title. II. Series.
CB478.R57   1993   303.48'3—dc20   93–18432 CIP

ISBN 0–8191–9090–X (cloth : alk. paper)

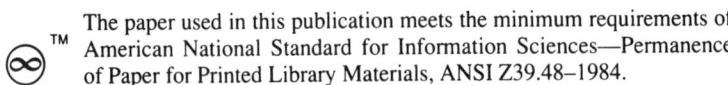

The paper used in this publication meets the minimum requirements of American National Standard for Information Sciences—Permanence of Paper for Printed Library Materials, ANSI Z39.48–1984.

L'homme est né libre, et partout il est dans les fers.

Rousseau, <u>Du contrat social</u>

Table of Contents

Chapter 1. Introduction: The Definition of Terms and Conditions 1

Chapter 2. The Phenomena of Technology: Essential Concepts and Fearless Misapprehensions 15

Chapter 3. *Termini Ad Quem*: The Limits of Technology 31

Chapter 4. The Point of No Return: Progress and the Linear View of History 43

Chapter 5. The Crossing of the Styx: Stability, Sterility and Death 55

Chapter 6. The Adulteration of Culture: The Impact of a Multitude 67

Chapter 7. Lost Among the Stars: The Secularization of Religion 83

Chapter 8. A Shortfall in Knowledge: Ignorance and the Proliferation of Information 93

Chapter 9. A Disparaging Condition: Challenges to the Self 105

Chapter 10. Human Bondage: Technology and a Technological Artifice 115

Chapter 1

Introduction: The Definition of Terms and Conditions

1

To live in the world without technology is inconceivable. In some way, shape or form, technology has always accompanied man, a fact signifying that we invariably see the world in need of alteration. Whether applied by means of stone tools or nuclear power plants, technology is evident among all peoples. Like religion, it is made manifest in all ages. But the present age is unlike any other because not only does it have technology, it also has a technological theme. Thus, we live with both the effects of technology and a technological view of life, enabling the very presence of technology to dominate our period of human culture. The present age has evolved into a time when techniques, technological progress, and a technological representation of life have given rise to a specific and peculiar morality, in fact, the predominant morality of our time.

The evolution of technology has a long and varied history, however slow-moving it was through all the millennia before the modern age. Its evolution was caused more by little contributions of innumerable individuals than by the chosen few innovators who are customarily accredited with all the inventive energy. Original innovators indeed are few, but their contributions are applied and sustained by all of us. From the simplest inventions of prehistoric man to our sophisticated, almost futuristic innovations, technology is a most captivating phenomenon. When we think about technology we ordinarily perceive it as changes made upon changes, modifications upon modifications, forever transforming the time-hallowed manners and mores of the past into the altered ways of the future. But how we perceive technology, how we believe it functions, may not agree with its reality. What will unfold here is no history; nevertheless this analysis is weighed down by historical awareness. And awareness, not so much of technology itself, but of technology's influence, has caused the present age, unlike all previous ages, to develop a philosophy of technology, a new philosophical pursuit of an old phenomenon, a phenomenon that attracts our attention and stimulates our curiosity, for it is one of the most powerful forces at our disposal. The groundwork for the philosophy of technology was laid in this century primarily by two thinkers: Ortega y Gasset and Heidegger; the former discussed the existential effects of technology, whereas the latter (although influenced by the concepts of yet another writer, Ernst Jünger) described its ontology. My concern parallels the attention Ortega y Gasset and Heidegger have paid to technology, since my emphasis is technology's cultural ramifications. One cannot write a philosophy of technology that stresses the latter's impact on value, without at least considering the thought of these two philosophers. Although there are many works which discuss technology from a purely analytical perspective, they add little to our understanding of technology's far-reaching impact in the modern age.

Mankind has developed a philosophy of technology because we require it, because the presence of technology as a most powerful force demands our close attention. But no philosophy of technology is written without some knowledge of its history, of its development and use over time, and the modern age is the first age to attempt an understanding of technology's values, to explain its place in the world, to formulate a comprehensive approach of its being. Despite sociological critiques of technology or purely analytical ones, both of which ignore the thought of Ortega y Gasset and Heidegger, my main concern, as was said above, is with technology's impact upon culture, upon the patterns of human action that are derived from the use of freedom, but conditioned by history. And culture shows it all--how we see ourselves, how we create a world which conforms with our perceptions of it, how we manifest love and hate, how we create ideas, moralities, religions, how we express admiration and fear, and how we fulfill basic needs and wants. Above all, it reveals to future generations what kind of men we were.

2

In order to arrive at a definition of technology, we must explore its etymological roots. To say that technology is a type of skill given over to means utilized to produce a practical end tells us little about the application of technology in our time. Etymologically, "technology" is derived from the Greek term *techne*, which signifies the embodiment of skill, a kind of discretion or faculty conditioned by circumstance. And the earliest example of *techne* that lends itself to criticism is a reference in Plato's *Protagoras* (319a) which comprises a discussion between Protagoras and Socrates about a specific type of skill that is essential to building good citizenship, in this case, a political *techne*. What Plato means by *techne* is attributed to the description offered by Protagoras in this dialogue, but Socrates offers a criticism that *techne* may be incapable of being taught. Since Socrates doubts if this type of *techne* is teachable, he refers not to knowledge, but to skill. Knowledge is teachable because that is how it is known, but individual skills might be unteachable, as the Socratic maieutic method of recollection demonstrates, because there may be no way to impart information that can bring out latent capabilities. We all have known people, for instance, whose language skills were simple and unrefined and were not improved no matter how hard they tried because they possessed no innate ability. Of course, other skills may be teachable, but the type of *techne* Socrates criticizes here may not be.

However, the classical definition of *techne*, like so many other words, has been challenged by Heidegger in his numerous contributions to technology; he interpreted this term not as skill, but as knowledge, contending that *techne* signifies a way or method of knowledge by revealing and therefore knowing beings, or, in Heidegger's manner of speaking, a bringing forth of beings out of concealment and into the unconcealment of their presence.[1] No matter how we define truth (*aletheia*), ultimately subjectivity plays a part because man plays a part, and this is true whether it coheres or corresponds to truth, whether or not it is performative or pragmatic. Regardless of Heidegger's unique definition of truth as the unhiddenness or uncovering of being, he assumes *techne* to be an element of truth because it is revealed through disclosure. In Heidegger's reasoning, *techne* is not the simple act of making beings by means of skill, but the more elaborate manifestation or process of the making of beings. Although this interpretation has some basis, it is largely unsupported by the nature of *techne*.

Because *techne* means art or skill, it must still in some way presuppose knowledge; that is, knowledge is embodied in skill as a subordinate idea through its manifestation, rather than skill embodied in knowledge. This is evident, for example, when we patronize the services of a person who possesses technical skills, such as a carpenter, physician or historian. We seek these professionals because we not only need their knowledge, we also need their skill in relationship to their knowledge. If these professionals were simply to rattle off their knowledge to us, they would hardly be justifying our understanding of their function, and would hardly justify the necessity that we compensate them for their services. Let us take the example of the historian. If he is employed as a teacher, his teaching may do little to demonstrate his skillful and experienced capacity as one schooled in the study of the past. He may give us his knowledge, but he may not be doing so in his capacity of a historian, since an educated plumber could do the same thing. But when the historian applies his knowledge and experience to an interpretation of the past, we encounter one who is skillful in analyzing it. Conversely, a teacher of history may be eminently qualified as an educator, but still be a poor and inadequate historian, that is, a poor and inadequate scholar, hardly empowered to interpret and analyze the past skillfully. Skill or art is the end why there is a need for knowledge, and knowledge is pursued in order to make one more skillful in the pursuit of art. Knowledge, according to Aristotle (*Metaphysics* 918a 24), belongs to skill because skill presupposes knowledge-- but we must challenge the view proposed by Heidegger who places too much emphasis upon knowledge in relationship to *techne*. In light of the modern age with its excessive dependence upon both science and technology, it is understandable how *techne* has been confused with knowledge.

Nevertheless, we may call *techne* technical knowledge in so far that it pertains to human activity in general, but knowledge in this sense should not be confused with science, since *techne* is limited in meaning, that is, in so far that it is not as universal as knowledge. Technical knowledge is definable in the sense of experience or skill, and the latter is what we mean when we infer technical efficiency. *Techne* is skill primarily, whereas science (*scientia*) is knowledge. *Techne* does not mean knowledge so much as the skillful use of this knowledge. Thus, *techne* really possesses two ideas--the embodiment of skill and the skillful use of knowledge. Since *techne* means skill or technique, it signifies the ability to do something, and the greater one's ability to do this, the more skill one has. We refer to the example of the historian above. Knowledge, on the other hand, is *episteme*, that is, facts and ideas in the sense of understanding or an acquaintance with something. Since *techne* is perfected only with knowledge, it is easy to confuse the generality of knowledge with its skillful application. And since *episteme* encompasses more than *techne* because *episteme* is a larger concept, one does what one knows, but doing is always limited to what is known. Here, too, this confusion is created by Heidegger who associates *aletheia*, the unconcealment of the concealed, with the idea of manifestation, that is, with the idea of a coming forth into being.

Since *aletheia* is an ontological situation (to be situated in truth), it cannot be identical with art or skill; that is, *techne* deals with being as it becomes known in the world through man's agency, whereas truth deals with the essence of being as it is in itself. And truth is incapable of doing anything itself. That is why it is man who acts, man who reveals being because he is in the world. Technology is really a manifestation of being because and only because man exists in relationship with the world which he alone has created. Hence, *aletheia* and *techne* ultimately are antithetical because *aletheia* concerns the being of being, however subjectively interpreted, while *techne* masks it in its particular way; and this concealment is a cause of much disorientation in our time. Such a difference between these terms is utterly fundamental to the nature of these concepts, and it is unfortunate that Heidegger confuses them.

Because *techne* is a means of making, that is, a manner of production, it denotes practical performance. And practical performance is what we commonly call technique, since the latter is the means by which we achieve an end, a manner of execution, a method of attainment. Since there are techniques of performing just about everything, they are variable, but there can never be technical ends to anything--only technical means. They vary not only because individuals make different choices regarding how to act, but also because of the diversity of applied skill itself. Thus, a technique is a manner of doing something; a method or process of action as distinguished from knowledge; an agency or instrument of action; a faculty or cause that produces an effect; whereas technology may be described as a systematic approach to the application of technique. Hence, it is apparent that technology, the embodiment of techniques, is pragmatic; it always intends to bring something into the world, to alter what is given, to change, theoretically at least, the possibilities of change. But it is also apparent that technology is the larger sphere into which the smaller sphere of techniques is subsumed. Techniques do not comprise technology as some would have it; rather, technology comprises techniques. All techniques are means to some end, but not all means are techniques, that is, not all means are technical. So nontechnical means reside outside of technology's influence, although they are subjected continually to technology's persuasive invitation to adopt its methodology.

Technology signifies a methodical treatment of art or skill (*techne* + *logia* = *technologia*); the method to which it is subjected is a structure, a form that is distinguishable from its material composition and that gives unity to the whole. This description incorporates the notion of a system, of a purposeful regularity emphasizing method and procedure. Technology may be described furthermore as a system of systematization; to describe this relationship another way we may say that the systematization of techniques constitutes technology. Systematization assumes an orderly approach to reality, which is most noticeable in Western technology from the industrial revolution to the present. Because technology is systematic, its definition conveys the idea of organization. The systematic or methodical aspect of technology is characterized by the predominance of means over ends--in fact, by a complete absence of ends, by the lack of hard and fast objectives, by an absence of any horizonally-directed goals--and so it may be said to be consumed totally by techniques, a characteristic that dominates our age.

3

Because our age is literally consumed by the idea of technology, we subordinate the products of technology, its so-called goods and services, to its method. Since we are more concerned with procedures, the products of technology have become secondary to its method, secondary to the development and perfection of technique. Contrary to the form of technology manifested in antiquity, which was always concerned with ends, that is, with production, modern technology is consumed primarily by means, that is, with action. We really do not care what we manufacture as long as we produce something, as long as the means of production keep us busy, as long as we are engaged in the process of engagement. Modern technology constitutes a self-perpetuating and continual system of systems, not only in production of goods and services, but in everything that we do. Since we envision no end to the perfection of techniques, we believe that there is no end to technology.

The absence of a final end to technology has some bearing upon science, particularly because science and technology are now irredeemably linked in the modern age, but of the two, technology is much older than science and is not the servant of it. Historically, technology had little influence on science as it originated with the Greeks. In fact, early science was more affiliated with magic than with technology, and only in the seventeenth century did science and technology intersect. Before this time, they were two worlds apart. That this dichotomy could exist seems unnatural to the modern age; nevertheless, science was nontechnical before 1650.

On the other hand, the systematization of technology is truly of recent vintage, although it had its roots in the Middle Ages--in the ideas of Robert Grosseteste and Roger Bacon, and in the inventions of Villard de Honnecourt and Peter of Maricourt, and in many other lesser or unknown individuals. These medieval contributors to the development of technology must be distinguished from the molders of science who lived in the modern age, particularly, Copernicus, Galileo, Descartes and Newton; but all of these individuals, both medieval and modern, helped to create the present techno-scientific age. Like many other cultures, the West used and later modified ideas and inventions not its own, many of them originating in China, notably, paper, moveable type, gunpowder and the compass. But only the West developed the scientific method, and only the West had an industrial revolution. All other cultures that utilize modern technology have adopted it from the West, and this is true both for those cultures that admit their obligations to the West and those that categorically deny them.

Technology can always exist in the absence of science or where there is little scientific tradition because there can always be technique without any science at all, as evident in many primitive societies; but science no longer can exist without technology, because technique has become the means by which science in the modern age is perfected. Our age makes manifest the fact that both science and technology profit from each other. Science enables technology to expand by increasing its tasks on the one hand, and technology enables science to become more precise on the other. But of the two, science profits more from technology than technology from science, because technology has become the avenue by which science is acquired, that is, technology has become the means to the perfection of science. It is technology which has made science so scientific in the last century. Since science has become technical far more than technology has become scientific, science has become a mode of technology and an agent of it. The broadmindedness of science has become dependent upon technology's methodology. By itself, science is careful and circumspect. It is repeatable, but not contingent. It does not know the whole truth, and therefore it is hesitant and tentative. And obviously science exists because of free inquiry. The destruction of the latter means the jeopardy of the former. Despite science's circumspection, technology allows science to develop its own method, that is, its own technique, yet technology possesses none of the caution so characteristic of science. Consequently, like all technique, scientific technique has become assertive and assured of itself, arrogant and pretentious. Because science is affiliated both with its principles and its results, both with its laboratories and its applications, it has quickly accepted a technological artifice, and this merger has given rise to a reciprocal relationship: science makes technology into a mechanism of power, and technology renders science impassioned and extreme.

In a manner of speaking, we must go one step beyond Husserl's *Crisis of European Sciences*, by questioning how technology has warped the validity of science and distorted the scientific method. Technology is an explosive phenomenon, manifesting the bursting forth of technique into a natural world devoid of method; it is a phenomenon whose ends are illusory, resolving themselves into no more than new means; it is continually evolving in order to have yet other means. Wholly preoccupied with the means to predisposed ends, its so-called ends foreshadow other means. As a result, technology, which is never satisfied with its present state of being and continually on the way to its replacement, becomes a perfectionist's fantasy. It is so consumed by its own means that ends have become anathema to it, and thus the meaning and even the possibility of its ends are lost to itself. As we shall see below, the absence of ends is a cause of much devastation, both to nature and to man.

Technology gives way to improved means for hitherto unimproved ends, but its means are always definite and clearly defined. Technology reaps no benefits from investing in uncertain and questionable enterprises. Although science may fulfill the objectives of technology, technology cannot afford to waste its time on the difficult quests of science. Technology proceeds as if it has no time to lose, with the urgency of the newborn wildebeest calf which must stand erect immediately and keep up with the moving herd. Like one who does not see or is not aware of more distant goals, technology always aims at what is close at hand. It is unwilling to commit itself to objective uncertainty because it cannot build a bridge over the chasm of doubt. And so as a result science has today become the servant of technology. It has lost sight of its ends and is now consumed primarily by means. The uncertainty of science has been replaced by the certainty of technology.

But the characterization of technology symbolizes more than merely a skilled worker, a craftsman or an artisan. Man the technician is more than the oversimplified description of *homo faber* because he is also a builder, a manufacturer, a fabricator, that is, a doer through the use of technology. Because we are doers--since we are more than rational or contemplative by nature--the being of human beings is to act. So technology comes naturally to us and is a result of necessity bearing down upon existence. Of course, we are much more dependent upon technology today than in the past, and this dependency, in addition to the proliferation of techniques, is especially evident in the use of tools which are designed for specific purposes to be used in the performance of a task, but may also be used for purposes other than those they were designed for. Of course, tools can become obsolete when they are no longer useful for anything. A tool is designed through technology, but its ultimate use is determined by its application, that is, it is we who devise what is a tool, how to use it, and when to replace it with another. The tool itself is an extension of the body. It enhances the power of the muscles and the dexterity of the hands. Tools give added dimensions to our bodies that would be impossible without them, thereby increasing control not otherwise available in nature.

Although there are hand tools and machine tools, neither of these should be confused with the machine which is a mechanically, electrically or electronically operated apparatus, such as an automobile, telephone or computer. Ultimately, there is little difference between hand tools and machine tools, but there is a major difference between tools and machines. Although a tool is intended to be used for the end to which it is designed, the tool user himself has become the means, the very utilization through which the machine operates. The tool user is the manipulator of a tool for a specific end, such as the use of a hammer to drive a nail, whereas the machine is the manipulator of the means, such as the set procedures a computer operator must learn in order to use a computer. Man is the master of tools, but the machine is the master of man. Man uses the tool as an end, but the machine uses man as the means. And because the hammer, for example, is designed to accomplish certain well-defined tasks, such as driving nails, it has changed very little since its inception, but the computer program is continually changing as a means in order to make use of a computer. In a sense, a machine is a self-working tool, a tool that works by itself, presupposing, of course, its source of energy. But our age is more preoccupied with sophisticated machines than those simple machines, such as the wheel, inclined plane or pulley, that also fit into this description. In fact, these simple machines, all of which are inventions of antiquity, or in the case of the wheel, prehistory, are now so unlike modern machines as to be merely elaborate tools.

As we said above, man uses the tool as an end, but the machine uses man as the means. And the machine enabled technology, which in one form or another was always with man, to evolve into a system. The machine, aided by other factors, such as the pragmatism of Western science or the progressive optimism of Christianity, helped over time to develop all manner of techniques. Therefore, a fundamental difference exists between the ability to make and use tools (*homo faber*) and the domination of this ability (technology as a system). Technology means more than just the use of tools because it signifies a systematic approach to being, a scheme that is devised and organized according to a definite plan, like a strategy of domination. In order for technology as a phenomenon to survive and perpetuate, a system must be contrived that contains its own logic, its own rules, its own theodicy. Although technology is a contrived approach to reality, it is uniform, predictable and malleable, and for a system based upon technology to come into the world, it must transform or destroy the world which existed before it, particularly any world which is substantially different from what will follow. Speaking with the voice of thunder, technology attacks everything that is nontechnological, although anything nontechnological cannot help but be attacked by technology, because anything is susceptible to technique, anything is liable to the influence of procedure, anything may be regulated. In short, the possibility of technology's presence destroys all other presences. Although many pretechnological functions still exist and will continue to exist in a technological future, they remain incongruous in the present, foreign and eccentric to a technological environment. These eccentricities include, among others, the art, religion and morality of a former age. Although infinite, technology is finite, that is, it is nearly endless in its possibilities, but limited by its logic. But, more importantly, technology is limited by the very conditions it creates, since each turn it takes on the road of development eliminates other turns, just as the choice of one thing signifies the suppression of another.

Because we have a tendency to think that technology is naturally given due to its omnipresence, we conclude that it must be innate in our being. This alleged innateness is presupposed by Heidegger, who perceives technology as an ontological concept, in short, that technology is a situation conditioned by our being.[2] Since any phenomenon which qualifies as ontological must, in some way, be indispensable to being, it must constitute a part of its most fundamental underpinning. Knowing this qualification, it is questionable why Heidegger assumes that technology is an ontological phenomenon because, if anything, technology is incapable of improving us ontologically. Have we become better because we now have automobiles, laser printers and weather satellites? The relationship has been reversed: that is, technology is no longer an aid in the perfection of being, but rather being is now an aid to the perfection of technology. Although it is difficult not to identify with one's lifelong occupation and hence difficult not to ascribe a purpose to it, Heidegger's preoccupation with ontology has obscured his vision to the extent that he has placed technology at the fundamental level of man's existence. Man plans and chooses. Technology is what is planned and chosen. Yes: man and technology seem to participate in a mutual relationship, but this relationship should not hinder our thinking. There is technology because man brings it into the world. Technology takes form through the application of man's freedom expressed through historical and cultural conditions, albeit limited by them, which include, among others, the religion, laws and mores of an age.

But what exactly is ontological in man anyway? Freedom, of course, but freedom is merely an openness to being. And freedom seems to be the only ontological concept. Incapable of evolution, freedom is simply an event. It is neither a characteristic nor a quality of being. It has no essence, no structure, and is not a part of any evolutionary process. Above all, it is not subject to necessity. And the latter is the key to why technology is not an ontological concept--if technology is anything at all, it is always subject to necessity. It is linked to circumstance. If ontologically based, technology and technological development would not vary as widely as they do over time and place. Since there are many different types of technologies, for which the most extreme type is our own, they are derivative not from being, but from being's free expression in the world. Technology is conditional. It is not inherent in being, but is an attribute to our being in the world. It is acquired, that is, it is derived from culture. It is not ontological, nor is it the result of heredity.

Technology is not ontologically necessary for us, although it is a facility by which we modify our state of being. Although superfluous, technology influences the manner by which we exist in the world by increasing possibilities in multifarious ways. And the need for technology tells us a great deal about ourselves. It tells us that the use of technology produces what cannot be found in nature, since what is not at hand we manufacture. Technology is a modification of nature, an imposition upon nature brought about by necessity. Technology also tells us that if we in any way have the ability to make changes in nature, to alter and modify it, we will do so. It tells us that we do not regard ourselves to be satisfied with our lot in the world, that we perceive the world as a prerequisite for change. Just as we are subject to change through the medium of time, we also live in the world of our own making. But at the very time when we are literally overwhelmed with technology, we are at a loss with what to do with much of it. We can understand technology, but often do not appreciate the consequences of its application. Popular mechanics is not for everyone. Like the Romans, we put technology to use, but do not theorize about it. That it works is our main concern.

Of course, man has always altered the conditions of nature. We try to make life better for ourselves from that life which nature gives us. From the use of fire and the invention of the wheel to the domestication of plants and animals, we try to improve what we find in nature. We make the artificial more to our liking than the natural. Thus, we have the ability of disengagement, because we have the ability to supersede the basic necessities of life by our detachment from those conditions which tie us to nature. It is not so much being but well-being that justifies Ortega y Gasset's definition of technology as "the necessity of necessities."[3] But not only does technology create needs, it also is created by them. It is the means used to fulfill needs as well as the self-perpetuating end created by necessity. No longer is technology based upon individual needs, but upon the imperatives of its own logic. In the present age, technology commands our all-encompassing attention. We are now driven by a deep and insatiable necessity, by a technology that plans both for today and tomorrow and is utilized for specific needs and long-term goals. It is concerned primarily with innovation, with its impact in today's world, because it must secure a footing in the present in order to have an influence in the future. In a sense, technology must be attuned to today's morality in order to affect the morality of tomorrow.

Technology improves our lot by enhancing the convenience of life, by making life easier, more comfortable, more dependent. Materially, it enriches us. Ontologically, however, it impoverishes us because it drains us of the source of our strength. Moreover, it covers up existence; it widens the gap between man and being. Like morality, technology does not help us in the least to choose. It is no aid to our use of freedom. It does not exist despite freedom, but only because of it. In fact, the conditions of the present age allow us to philosophize about technology--for which all previous ages had no need. Although the task of philosophy is to help us understand ourselves, to disclose being as a means to know ourselves, technology obscures this task because it obscures human reality. Since technology technologizes every aspect of life, it is guilty of making our assessment of ourselves increasingly difficult. Yet, ironically, the increased use of technology makes the pursuit of traditional, humanistic philosophy that much more indispensable, regardless of technology's open and arrogant disregard of this pursuit.

As we have already said, there is no technology without technique. Nevertheless, technique alone is not the only prerequisite for technology. Technology presupposes a morality, an underlying value that determines its impact in the world. Modern technology embodies more than a single moral concept; it has become the predominant morality of our time. More than an ideology, it has been transformed into a way of life. It must not be considered merely in its effect as a morality; whereas morality is always projected toward some end, the end of technology is forever more technique, that is, unending increase in its impact as a means, and ever-continuing augmentation of its influence in the world. This is its moral imperative. Thus, technology is never neutral; it is either good or evil. The claim of neutrality is merely an attempt to divorce technology from culpability. Even the claim that morality has not kept pace with the progress of technology has no foundation; technology itself is a form of morality.

Since means are morally relative, the means of technology pushes this idea to its limits. Ideally, we can say that anything in itself does not embody value. Rather, value is embodied in things based upon moral choices. However, certain things already embody particular values because of what they are and because of the end or purpose to which they apply. Since technology is man-made, its very creation presupposes a value that accompanies it. Therefore, technology is not innately neutral. If there are certain technologies that are innately good, then there conceivably are also technologies that are innately evil. There is a limit beyond which technology becomes harmful, that is, there is a threshold that divides the careful use of technology from reckless endangerment. Once this threshold is crossed, we lose control of technology or, to be more precise, we lose control of ourselves as masters of technology. Since we are the driving force behind technology, we are responsible for what it does as for what it does not. We may find scapegoats to blame for our failures, but the failures of technology ultimately fool no one. If actions speak louder than words, technology's actions belie its claims. In fact, the techniques of technology always presuppose intention. We cannot have a method or a process without also having the means to apply it, that is, an attitude toward its use in the world; for intentionality, as indicated by Husserl, implies a purpose toward attainment. And our purpose is domination.

Introduction 13

Notes

1. Heidegger, "The Origin of the Work of Art," in *Poetry, Language, Thought*, trans. Albert Hofstadter (New York, 1975), p. 59 [Translation of "Der Ursprung des Kunstwerkes," in *Holzwege* (Frankfurt am Main, 1950)]. Cf. Heidegger, *An Introduction to Metaphysics*, trans. Ralph Manheim (New Haven, 1959), p. 159.

2. Heidegger, *The Question Concerning Technology and Other Essays*, trans. William Lovitt (New York, 1977), pp. 21-22.

3. Ortega y Gasset, "Man the Technician," in *Toward a Philosophy of History* (New York, 1941), p. 99.

Chapter 2

The Phenomena of Technology: Essential Concepts
and Fearless Misapprehensions

Since we are in an open relationship with being, we are able to perceive our own being in contradiction to all others. We are able to observe the relationship between ourselves and the world at large. Thus, we are empowered not only with self-perception but also with the faculty of standing back from the world and observing it abstractly. Because we can mentally separate ourselves from the world, we can hypothecate a world different from the one we see. We can consider a world with improvements, that is, we can make modifications upon nature for immediate goals; we can innovate. The impact of that innovation is a slow and methodical change utilized for specific needs, rather than for long-term effects. Only with a footing in the present can innovation influence the future. Such an ability is the means by which we introduce change into the world, for change is an alteration of what is given.

Although there are many concepts that manifest the phenomenon of technology, there is none more evident than change and adaptation; change is that revision introduced into the world by our manipulation of the natural environment, and adaptation is its assimilation. Technology is the result of reason conditioned by the weakness of our own bodies and by the harshness imposed upon us by the physical world. A perfect being in a perfect world would have no use for technology. Only a being subject to choice and imperfections would stress the necessity of modifying nature. Since it is inconceivable that we, being free, would ignore the options made available to us, we have always felt this necessity. By all means at our disposal, we have always taken the opportunity to modify the world when necessity presented itself. Thus, technology dawned with the dawn of man.

Technology accompanies us wherever we go; it has always been with us, but this omnipresence does not signify that technology is ontologically derived. Although it is true that we have always modified the world, modification is not the result of technology being innate in man, but the result of the world bearing down upon us; we create and impose change upon the world because we see a need for it and we adapt the changes we introduce. Technology exists because of human culture--that is, man influenced by custom, yet conditioned by temporality, chooses to modify what nature provides. If technology were a quality of being, that is, if it were ontological, it would give us no alternative because we would be eternally subjected to technology and its methodology, since we would be condemned to use it regardless of free choice. Such a constricting point of view is indeed a prime cause of our troubles because it deludes us of the necessity of freedom when saying no. Because technology is so much a part of our world, it is nearly impossible for us to envision a world without it.

In point of fact, technology would not exist were it not for reason, that faculty of intentional and careful deliberation by which we view the world in an abstracted manner. Notwithstanding, the only insight technology possesses of itself is its rationality, that is, its methodology. Apart from this one insight, technology is devoid of all others. Although reason gives technology a purpose, in fact, a single-minded purpose, it cannot give it will or impulse or instinct. Certainly, we can have all these feelings for technology, but technology in itself is not characterized by any intensity of feeling. Rationality's aloofness cannot give technology that force which would direct it beyond itself. Technology is solely a self-perpetuating means because it brings no final satisfaction to anything it does, nor can technology ever become an end in itself because there is no end to our need to modify nature, no necessity to end our control.

Technology's use of reason is domineering; it is founded upon technology's omnipresence, upon its persistence in penetrating every place into which reason takes it. Even irrationality becomes consumed by technology's rationality, that is, rationality incorporates the free, and sometimes irrational, expression of creative ideas. Thus, the scientific laboratory becomes a prison because free experimentation succumbs to the excesses of scientific rationality. Because of this, technology is too much with us. It is forever present, like the moaning of the wind. It leaves nothing to chance, nothing to uncertainty, nothing to nothingness. It leaves no stone unturned. Its ubiquity is simultaneously vertical and horizontal, reaching out to everything, making contact with the world at large. Technology has invaded every corner of the globe, from tropical jungles to snow-covered mountains, from the equator to the poles. No longer can we hide from technology because, like the rain itself, it falls upon all of us. And like a monsoon, technology floods the land, drowning all by its presence, while for those who wish to flee, there is less ground to stand on. Every niche and every crevice fall to its influence, and with it also fall forever many plant and animal species.

This continuous and ever-expanding technological presence gives us the feeling of having lost the sense of direction. Although alternate choices are often available, the development of technology is characteristically linear. Thus, technology's ubiquity annihilates all primitive cultures. It obliterates all nontechnical languages. It challenges indeterminate ends, uncertain consequences, discontinuous manners, indiscreet features and fleeting glimpses of obscure worlds. It acts with cold indifference to the plight of elemental and natural societies. The whole world has become westernized, not because of the superiority of Western culture per se, but because of the superiority of one aspect of it, its technology. It is that to which the world gives its greatest aspirations and its highest hopes. More than Christian missionaries, Western technology has colonized the world; it is the new imperialism. Indeed, our best salesman abroad is not our literature or religion or art, but our technology.

And technology has become the most efficient system because it is the most rational. Its rationality compels efficiency, overwhelming whatever is nontechnical. Because of this influence, technical efficiency is productive to the point of nausea. Since technology compels agreement by the pure force of reason, it prepares the way for a complete subjugation of our activities to a technical rationality, and to the ingrained expectation of some accredited procedure, so when we travel to the far corners of the earth we expect to meet the same efficiency we are accustomed to at home. Just as freedom is expressed primarily through negation, so too is technology, that is, technology's affirmation is the negation of everything else. If we are to live in a technological world, we have no choice but to say yes to technology, that is, we must never say no. We have made ourselves technology's agent, its efficient cause.

Because we feel that we can only affirm technology, we make ourselves vulnerable to everything it does. We become susceptible to its exploitations, its shortcomings, and its failures. Like the Spanish conquistadors, we do not question the purpose of what we do because we have no doubt about the validity underlying our actions. We conquer with the same assurance that we can do no wrong. And like these conquistadors, the cries of the vanquished who believe in a discriminating technology cannot move us. When we are blinded in this way, the affirmation of technology outstrips the capacity for its rational use and we are forced irrationally to be hounded by reason, so to speak. In time, a world that surrenders itself so readily to technology is cause for a new form of brutality, since rule by reason alone is a form of oppression. But if we question reason, we cannot ignore its critics, for if we find fault with Voltaire, we must do likewise with Maistre.

The use of reason is most evident in the computer, in that electronic machine processing information by means of a predetermined program. Because the computer mimics human thought processes, however simplistically, whenever the computer is available, it quickly follows upon the heels of man. Thus, wherever we go, the computer invades, if for no other reason than because it is a machine that is modeled upon logic, and logic is parallel to reason. Although reason is often regarded to be our principal characteristic, we can also be sensual, irrational, emotional. Indeed, if we are not totally rational, then the computer will always be out of step with us because it is impossible for a machine founded upon logic and mathematics to act deliberately in an irrational manner. Thus, the computer ultimately will always be a little alien to us, a little anti-human, unless we give ourselves in totally to reason. The computer, when we allow it to rule, is representative of the machine at its worst--incapable of rendering judgments, unable to weigh conflicting possibilities, consumed completely with quantitative analysis. These traits also epitomize us when we are in less control of ourselves.

It is not surprising, although worthy of ridicule, to hear the computer described as a psychological machine--a machine with a soul, a psyche. Such a misconception is both typical and incisive--typical because it is expressed far too often, incisive because it reveals a gross ignorance in our thinking about the computer. We project our highest ideals, which are technological, upon the computer, until technology becomes everything and we become nothing. It is childish to imagine that computers have a brain; they only have a memory, nor do millions of data bits stored away qualify as thought. When it is said that computers think, what is meant is not that they engage in reflection, but that they call up and process particular things. Moreover, the "thinking" of computers is inexorably logical. If we ever get them to behave illogically or to forget, then we will have really developed something. Nevertheless, computers have been credited with a value far beyond their merit. They are even considered by some to be more akin to us than animals, hitherto, our closest sentient beings. But even as imperfect as we might think animals to be, they have no need of technology. Their perfection is atechnical.

In its infallible consistency lies the computer's great attraction and the reason we set it up as the ideal. We mimic it; we strive to be as predictable as it is. We attribute a greater perfection to it than we ourselves possess. But the computer's greatest attraction is that its logic is democratic. The computer is the universal machine of an egalitarian and civilized world, and it permits anyone to use it. It is the great equalizer, requiring neither unique talents, nor special skills, nor moral preference, nor acute wisdom. It is devised for anyone and everyone. It is the machine par excellence for the masses.

The computer quantifies reality, and in so doing it becomes the ultimate invention of a technological age. It objectifies everything it touches; it reduces reality to its simplest characteristics by numerating everything. Of course technology in general has predisposed us to quantification--nevertheless, it cannot quantify subjects unless they are objects to begin with, which is a way of saying that people are replaceable by machines when they are treated as such. But technology merely displays the human qualities it begins with, that is, if we are astute and circumspect, technology will be also, but technology will be reckless and threatening if that is our nature. Technology can only intensify human qualities; it cannot create them. If technology is obsessed with power, it means we must be so prior to it. If technology exerts absolute control over nature, it means that we desire the same manifestation over others.

These distinctions should help us avoid the temptation of conveying qualities in quantitative terms. Since quality is indeterminate, that is, nondeterminate because it is grounded in freedom, quality is a revelation of being. Quantities can be subdivided into smaller and smaller components; qualities cannot be separated from the whole and still remain intact because qualities are attributes reflective of a unity, that is, the qualities of anything are meaningless unless they are applied to individual entities. Nevertheless, technology, and most notably the computer, deal only with quantitative data. They can only manipulate those characteristics that they can know. Qualitative data is worthless to technology as noncomputable data is worthless to a computer. This is why a technological view of the world, which is much taken up with things and little with metaphysics, demonstrates the triumph of Hume's empiricism and its obsession with quantification; natural phenomena is quantifiable but being is not. Thus, the modern analytic view of the world introduced by Wittgenstein, which reduces everything to the smallest components, to the most rudimentary basis for any assessment of the world, fits well with this description of the quantification of reality. Understandably, this view is very limited because it restricts itself to minuscule segments of non-human reality. What Wittgenstein's analytics tells us about the world is very little, and what it tells us about ourselves is nothing at all.

Quantification distorts the perception of reality because quantity becomes confused with quality, although quantity in itself gives no value to what is quantified, and when that happens, we encounter substitution, or to be more precise, mediocrity--the condition where one entity is considered to be as good as another. Understandably, any increase in quantity, such as in population, causes a decrease in quality in all those individuals that make up the mass. Likewise, a reduction in numbers may signify an increase in quality. Where quantity is regrettably accepted as a substitute for quality, it acts to our detriment--an increase in size does not guarantee an increase in wealth, choice or strength.

One may object that quantification gives to technology the power to equalize everything, but it performs equalization on merely a small scale.

Equalization makes everything small. This is why technology goes hand in hand with a leveling process, and the more mediocre the principle of technology, the more necessary it is to perpetuate its importance. The leveling effect of technology most notably has destroyed the hope of producing anything exceptional. It has also destroyed the milieu that was receptive to the exceptional. Technology's leveling effect confers the same "distinction" upon everything and everyone, so that distinction loses its significance. For there to be distinction, there must be discrimination. Technology is promiscuous, that is, indiscriminate: when every color is gray, no color is distinguishable. The absence of discrimination means that technology accepts everything, including what we hold dear and sacred, as well as what we hold in abhorrence, so long as it promotes its own advancement.

Much like democracy, technology does not demolish opposing views so much as incorporate them into a larger structure. It accepts free thought as long as it leads to practical applications; it tolerates individuals as long as they enhance progress; it endures differences of opinion provided that they are methodical; it favors all sorts of oppositions only if technology itself is not contradicted. Technology will do anything it has to in order to survive, but underneath its appearance of tolerance--like Christianity--is an intolerant, if fashionable orthodoxy. It creates the impression that it liberates us, that it enables us to accomplish more with its aid than without it. But this is a delusion because although technology enhances possibilities on the one hand, it limits them on the other. As quantity increases, technology limits the possibility of quality, and this relationship holds for both the quality of life and the quality of manufactured goods. But a most significant example how technology increases possibilities on the one hand while decreasing them on the other is the relationship of man himself to both nature and technology, for technology increases the possibility that man will be the master of nature, but it also increases the possibility that technology will be the master of man. And more than making us dependent upon technology's methodology, whose effect in itself is far-reaching, it makes us totally dependent in our very thinking, in our orientation to problem solving, in our view of life. I should also add that we are potentially most ignorant of the impact of technology at the very time when we are most assured that we understand it.

The calculated, analytic and formalistic view to which technology has reduced the world's complexity is poor in meaning because it reduces the essential to mere superficiality. Such mechanization cannot help us in finding solutions to human problems because it renders the world cold and distant. Since it reinforces quantification, humanity in general and individuals in particular have become fragmented, cut off from each other, divided among themselves. Mechanization is the very organization of technology, so that as the whole world becomes increasingly similar, we have a greater tendency to become trite, banal and commonplace in everything that we do. Certainly the last thing that would result from mechanization is the development of a critical, acute and refined discrimination. Even when we act with sophisticated machinery and elaborate techniques, we are more primitive, less artistic and less refined than our predecessors. Like the modern age itself, we have become flat, having no depth, lacking in vigor, denuded of both shadows and contours, peaks and depressions; we are dull.

The modern age is not the first age to view man mechanically, nor is it the first to take an overall perspective of mechanization. Attraction to the mechanical is an attraction to our own inventiveness, to anything that is man-made. In a sense, such attraction is a form of self-love. It is also a desire to be God, to bring things into the world, to assign them a value and to replace those things devalued. Therefore, it comes as no surprise that we confer human qualities upon machines and ascribe mechanical qualities to ourselves. We have even attributed mechanical qualities to animals. Although this characterization is evident, for example, in Cartesianism,[1] it should not be equated with the unthinking, nonreflective automatons advocated by Pascal, who like Montaigne, believed reflection to be dangerous to man's development. Nevertheless, we share physical conditions with animals, but they do not share the existential condition with us. They are what they are without possibilities because they are complete. There is no deficiency between an animal and its being because there is no need for becoming.

Because mechanization is easily accepted as the substitution for authenticity, imagination or spontaneity, mechanization treats people as things, regarding them merely an fabrications of its design. This occurs not so much because technology appears unnaturally in the design of things, but because it passes off so-called artificial substitutes as intrinsic products of progress and regards its own effects to be natural while regarding nature to be an imposition to technology's development.

But the greatest degradation that the modern age experiences through the mechanization of technology is warfare, which has produced war on the largest scale the world has ever seen. Unlike wars of the past, which were fought primarily for limited purposes, localized to a certain territory, requiring just the force demanded by circumstance, modern wars are now global in scope, not limited to a specific terrain, no longer conditioned by force derived solely from circumstance. And these differences are due largely to technology. Modern warfare represents the full-scale mechanization of both men and material, and demonstrates that preparation for war is largely technical. Because not only armed forces but civilians are thought hostile, mechanized warfare intends to destroy the enemy and terrorize civilians. It obscures the differences between combatant and noncombatant.

Whether utilized in peace or war, technology is always a form of power. Since the power of technology increases with the spread of its dependents, it becomes everything to everyone as a means to heighten its necessity. The form of power associated with technology implies more than just control over nature. It also implies control over man and over the use of technology itself in the future. Thus, the purpose of any system founded upon technology, devised specifically for everything and everyone is domination, whose purpose is the control of all that is. But power is never an end in itself, which is especially true for technology. All forms of power are simply means to action, that is, means to control. This is why the state plays a principal role in the development of technology, and why its use of power is always self-serving. The political use of technology means the centralization of power, above all, the concentration of power in the hands of a few. Whether the latter are called politicians or technicians, they do not need to be particularly beneficent, cultured or good. They only need to be powerful. Cruel, ignorant and evil men could just as easily possess this power. In fact, such men are found in all mass societies, so-called egalitarian states, where the latter have the potential to evolve into one of the worst forms of government because they develop into a most efficient system of domination. Thus, the use of technology by the state becomes a tool of oppression so that all states in a technological age retain the latent potentiality of becoming totalitarian.

Power may be centralized, but knowledge cannot be. This is why it is easier to use the power of technology than to understand it, and easy to suffer from its ill use. Power and knowledge are not really coterminous; they are often opposites, and an increase in power may ultimately lead to a loss in knowledge. Knowledge sometimes stifles action, that is, the type of action that leads to power. So we should disregard the popular belief that it is not possible for power to be exercised without knowledge. When Bacon said that knowledge is power,[2] what he meant was that knowledge enables us to do something because we know how to do it, and the doing of it in turn makes us powerful. But power is not so much created as it is discovered. We take it seriously because we have a need for it or because we find little hope of avoiding its presence.

If the growth of power destroys our capabilities to act wisely, then technology prevents us from making sound judgments. The very presence of technology compels us to use it, even if it is self-destructive. Hence, we pay allegiance to technology in order to become a part of its power, even if the use of this power is unsound. Technology seeks predictability, that is, inflexibility, so although it wants us open-minded in the beginning, it insists that we be closed-minded thereafter. It wants us clearheaded, nonintoxicated, sober, and after it has our attention, it then wants us drunk with it. Whereas in former times technology was a means to knowledge, with knowledge as an end in itself, allowing it to be perceived as infinite, unconditional and all-encompassing, in the modern age, technology is always a means to power, and power is finite, conditional and limited. Although knowledge and power are not coterminous, the perfection of technology is merely the means to greater power, that is, the means to more control. Never is technology regarded as an end in itself. The tremendous advances technology has made in the modern world are grounded in this obsession with control.

When in the presence of technology, we always perceive the world as a cause for change, since it is not our nature to accept the world at face value. We seem to be driven by some force that compels us to alter whatever is given. Although we are not the only beings who modify our environment, since animals do this as well, they do not do it to the extent that we do it. We go beyond the basic requirements for food, clothing and shelter. We so alter the world as to create virtually a new world out of nature. Technology, as defined by Heidegger, is a revelation of our ability to disclose what is hidden,[3] to find and reveal the latent capabilities of nature, but this ability is actually the power to transform what is given, to change what is at hand. In itself, technology is incapable of concealment, even if one attempted it. And with the openness of technology goes the openness of change. Wherever technology secures a stronghold, change will also be embedded. But technology's changes are not continuous and unending; rather, its mode of change initially creates new structures and new organizations which then solidify, rejecting any additional changes, especially any changes that conflict with technology's initial changes. Although confronted with new products and new models of manufactured goods, although subjected to new forms of bureaucracy and social control, we will never see revolutionary change, that is, we will never see accelerated social changes which are the true changes of any revolution.

Although technological innovation gives novelty a value, change in itself is valueless, especially when change is substituted for substance--as is apparent when change is introduced for its own sake, with the initial result that it is merely a catalyst to other changes--but only in so far that these changes are executed within the methodology prescribed by technology's systematization. We must keep in mind that technology is concerned with what we change, not that we change. Since mere change is highly uncertain, in fact, unpredictable and potentially dangerous, this type of change is most threatening to technology, since these changes challenge technology's methodology. Indeed, a technological age is not in the least transitory even though it strives to be both current and fashionable. It is an age that produces nothing lasting, marked by ideas which have no chance of introducing truly meaningful changes into the world. The changes that technology brings induce us to think of it as motion, continually evolving, constantly flowing, but actually we go nowhere, except around in circles. Technology emphasizes speed. It reveres rushing about as a supreme characteristic, even if we are going nowhere in particular. Although theoretically directed to the future, to the unknown and as yet unknowable, to the nebulous void of time without form or purpose, technology's main concern is actually with the present where everything is discernible, manageable and controllable.

Need plays a part in the technological scheme of things because need is both a cause of technology and an effect of technological change. Need stimulates creative effort, that is, necessity is presumed to be the mother of invention, although one can have inventions before there are needs for them. Need is a stimulus to technology just as technology creates needs that do not yet exist, that are not yet felt, that are not tied to the moment. Of course, technology does more than just create needs that do not exist; it also perpetuates needs that already exist. In fact, we now want things that we never knew about before, much less needed, but we also want them because other people already have them. Technology creates needs that come into being all by themselves, needs that make us pursuers of mere externals. It gives us a feeling of exigency, resulting in the profound and personal craving for the profusion of meaningless and trivial goods and services fathered by our age. The needs we have created by reason of technology should more accurately be labeled wants, but we have also concealed their technological derivation by relabeling wants as needs. So much so, genuine and biologically derived needs, which are limited, such as the need for food, clothing and shelter, have now expanded to incorporate the notion of technologically disguised "needs," that is, wants, which are unlimited. These would not be so bad in themselves were it not for the fact that they also remove us from the fundamental.

The more technological the world becomes, the more we expect to be surrounded by the products of technological development. In fact, the absence of these products makes us uncomfortable, since all of us have grown so accustomed to them. In addition to adapting ourselves to their presence, we are also encouraged to acquire and utilize these products, the very goods of industrialization. Consumerism, therefore, perfected and encouraged by technology, makes people self-protective, materialistic and petty. It excites avarice in people whose appetites are conditioned by things so that consumerism becomes the handmaiden of technology. Because one goes hand in hand with the other, we become familiar very readily with acquiring and owning things because the world is so full of them. So in place of doing something meaningful, we become consumers. In our world, things have taken the place of words, and not just any words, but words of understanding. Increased consumption is seen as the road to plenty, making shopping our greatest pastime, which helps us to forget our troubles by seeking entertainment or social status in the absence of real, meaningful activity. Since the lifeline of technology is based upon consumption, its decline forestalls economic ruin. The consumption of industry's goods and services is transformed into an end, that is, consumption becomes a reason for technology. The products of technology which have little or no lasting value in themselves are manufactured and consumed merely to enable us to persevere from day to day. Consumerism becomes a perpetuating system, driven by the fear of poverty. Enticed and cajoled by advertising, our modern designation is no longer citizen, but consumer; one who does not determine what is offered for sale, but merely determines what he consumes and buys what industry determines he will buy.

Consumerism is valuable only if it is continual. It presumes that we must discard old and outmoded objects for new so that goods are no longer valuable by their use, but by their replacement. This condition renders materialism impractical because the utility of things is expressed not by itself but by what follows it. Thus, we have developed materialism with little regard for the usefulness of things with an insatiable appetite for renewal of the old by the new. We have brought into being an impractical age based upon continual dissatisfaction, not the dissatisfaction we may have because we have not yet achieved the perfection of ourselves, but because we have not perfected the last perfection of things. And innate within consumption lies waste, which results because manufactured goods are intentionally produced and deliberately purchased with the idea that after a short duration they will be discarded. No product lasts for very long because longevity is contradictory to the needs of production.

Hence, convenience is equated with modernity. In point of fact, technology strives to make us comfortable; it seeks to make everything pleasurable. Yet it gives us no real basis for pleasure. But more importantly, it is to technology's self-interest to keep us satisfied because dissatisfied people cause problems. We can be just as easily controlled by pleasure, or corrupted by it, as by the fear of pain, because the convenience brought to us by progress strives to curtail life's hardships. And when corrupted by pleasure, we lose more than the ability to sacrifice; we also lose the ability to share because our affective ties have been weakened. Misfortunes bring us together; dissipations pull us apart. But technology, which attempts to remove all obstacles to inconvenience, all barriers to discomfort, all impediments to suffering, makes us complacent. And when we feel that we have fulfilled ourselves, that no more challenges exist for us, we have entered a period of negation. It is when we are a little restless, a little dissatisfied with ourselves that we still care for life, that we can profit from struggle. Like a fine cutting tool, we retain a keen mind and a well-tensioned body only with constant grinding on the great stone of work and toil. When we are no longer sharpened, we dull. Hence, the lack of hardship very often becomes an obstacle to greatness just as pain frequently is the means to creativity.

The assimilation of pleasure also causes laziness. This is because technology makes everything easy, which goes a long way to making us indolent. And indolence makes us spectators, not participants. Hence, we become observers, like spectators at a ball game. The comforts of technology have only made us more expectant of further comforts. We now expect a higher standard of living, a greater stability and a more homogeneous environment, all of which are dependent upon technological development. Laziness must be distinguished from idleness: laziness means that we want to do nothing or very little at all, whereas idleness means we want to be inactive and unscheduled for the present, but anticipate that activity will be engaged again at some point in time. Here, too, technology plays a role because on the one hand it makes us lazy, on the other it discourages idleness. The use of technology makes us totally dependent upon it, totally a spectator, totally tied to its methodology. Even the free time or leisure that we receive from technology is not used to perfect ourselves, but to waste time in meaningless diversions, thereby dissipating our energy. We do not live so much but merely pass time, like waiting to die. So we pursue sports rather than the love of wisdom.

It follows from this that we have become accustomed too easily to immersing our senses in the noise generated by technology, thereby making it difficult to listen to ourselves. We must resist the flood of mass media, we must stand back and see ourselves apart from the many, we must identify ourselves from all other selves and be internally directed. Externally we find only noise. Frightfully, we no longer have the feeling of being alone with ourselves. For not only are we bombarded daily by the noise of technology, we also are increasingly incapable of doing anything without its presence. We live in a noisy and disharmonious age, an age characterized by the deprivation of quietude, by the absence of that serenity we need in order to be at peace with ourselves and the world. If silence is the prerequisite to self-awareness, then the constant noise created by technology will arrest this development; it will prevent the self from ever being in its own presence. We must learn again to hear the silence, for even in a crowd, which is characteristically demanding and unrelenting, the silence of the individual is eloquent.

Indeed, technology looks upon solitude as its rival. But it is not so much that technology has little respect for solitude and no regard for its necessity in the perfection of the individual; rather, it regards solitude to be alien, even dangerous, because it removes the individual from technology's omnipresence. And therefore, technology makes the individual doubt his own individuality just as the master makes the slave fearful because the master fears him. Technology strives to be everything to everyone. It wishes to be considered whatever we do, wherever we go. It tries to fit perfectly with its surroundings, like the union of mortise and tenon. But technology also abhors a vacuum. It allows no exceptions and offers no amnesty. Always loud and boisterous, technology never whispers. Wherever it is, it shouts, because it is intended to drown out free thought. For these reasons, technology intrudes upon our privacy. It is always insensitive to the plight of the solitary individual, bombarding him with the noise generated by mass media. Technology encourages the needless proliferation of media that could not otherwise exist. Although the present age is deluged by the desire to communicate, it precludes the fact that there is anything to say.

And because a technological age appeals to ordinary people, it becomes an age of ordinary values. It is an age of small things, when neither the heights nor the depths of human endeavor are of much concern. But there is no contradiction here between the smallness of technology's values and the vastness of its influence. In fact, it is because technology has an interest in everything small that it is found everywhere. A technological age is truly an age of the common man--and what makes him common is the manner by which he views himself, the values by which he lives. These values are limited and limiting. They constrict his world to a smaller and smaller sphere of influence, despite the development of a global economy and international communications, both of which seem to prove that the world is getting larger. To put it simply, ordinary people live in a smaller, but louder world--smaller because this is their manner of operation, and louder because noise obliterates thought, consciousness and freedom. A loud and constricted world of ordinary people is noted for its triteness. It lends itself to pettiness. It is an age characterized by banality, since its horizonal quality or breadth of vision is narrow.

Since aristocratic values deal with the best that we can attain, a technological age conflicts with these values, or to say this from technology's point of view, aristocratic values are antagonistic to technology's. In fact, an age conditioned by technology is fundamentally anti-aristocratic, because it cannot afford to make exceptions. It must deal with the masses--mass media, mass education, mass morality. And what appeals to everyone in a technological age is ordinary--unvaried and indiscriminate values, values that no one would take offense with, values that are safe to the uncritical eye. In order to offend no one, these values must be harmless. They must be so empty of substance as to be ethereal, in fact so ethereal as to be almost nonexistent. The ordinary values of commonplaceness and triviality characterize an age that is no longer great. Important things are still done in such a time, but not exemplary things.

Another misapprehension of technology is that it is divorced from mythology. Myths have effect when we give them credence, ignorant that they are myths; they appear as myths only when called into question. Not only are myths frequently at odds with reality, but they often make it difficult to ascertain what is real, and one such myth is that a technological age cannot function if it is pessimistic. It must blindly and boldly forge ahead with self-assurance. It must be blatantly optimistic. Moreover, the notion of progress that accompanies our age is also permeated with excessive optimism, a notion which says that we can do anything. Progress implies that the world is becoming more comfortable, more humane, more prosperous, less susceptible to disease, ignorance and want. It also implies that natural resources are limitless, that we are limited neither by the sources of energy nor by nature itself which defines the physical bounds of the world at large. We have come to regard progress as inevitable, and that is why it is shrouded with optimism. But at all costs, we must avoid nurturing an overly jubilant and groundless optimism that lacks foundation and buries life's shortcomings, and we must resist the myth that progress relieves our suffering, that technology is the best anodyne known to man. A realistic look at technology makes it all too evident that, in order to be productive, technology must promise more than it can fulfill. In the end, it cannot mitigate the suffering of our existence, nor can it put an end to our mortality.

Technology also creates the myth that our age is better because we are totally given over to technique--that we are highly sophisticated, that we have become refined because we have attained such an acute level of technical expertise. But Western civilization was not founded by modern man, and owes little to his innovations. Philosophy and history as we know them, and above all, science and mathematics, were all created by the Greeks, who were primitive indeed by our standards. We simply maintain and extend their civilization which has lasted now well over two millennia. As was noted in the twelfth century, we see farther simply because we sit on the shoulders of giants. The world is a vast place, much larger than we usually envision, but it is not getting smaller; it is getting bigger. It is technology that gives us the illusion that the size of the world has shrunk to the distance it takes to travel between two points, that we are always in control of things, that we have fathered the present without the assistance of the past. But technology is grounded in more than scientific perception; it is delusive because it is always at odds with the senses.

Ironically, technology is never seen as a threat, never posed as a danger, nor as subjecting us to risk. But not only are we blind to its shortcomings, we are also oblivious to its authoritarianism. Technology refuses to have its truth put to the test; it tolerates no challenges to its authority. It portrays itself as the divine *logos* serving as the measure of all things, and we treat it so because we do not do anything without consulting it first. We do so regardless of its deceptions, whereby many problems caused by technology are disguised as opportunities to produce even more technology. Nevertheless, there is no irony in these contradictions, in technology's adaptation, in its unique ability to be abused by any culture, no matter how unsophisticated. Because technology spreads by means of information, it is applicable to any culture which can acquire and exploit it.

Technology's powers of assimilation blind us to self-individuation, which results in the betrayal of culture, in an alignment of the foreign at the expense of the familiar. And thus, the cause of our disarray lies deep. Since we have lost the ability to differentiate the genuine from the spurious, we have forgotten that the necessary condition to self-individuation is self-understanding. Because we are unaware of our true possibilities, we see our destiny as tied to the problems of the rest of the world. So we make it a point to meddle in the affairs of other nations. We also tell ourselves that technology has unified the world and all people in it, that it has created an apolitical multiculturalism--but in fact, the world is united only superficially; we cannot speak of one voice unified.

Although technology imposes a type of uniformity upon diversified cultures and societies, it achieves this uniformity at the expense of the infinite diversity of the human experience. This uniformity, which destroys the superficial differences among cultures, sensitizes individuals to different cultures but ultimately divides them, since differences, not similarities, are more evident. The trend toward a universal brotherhood is halted because uniformity brings prejudice, that is, it brings awareness of those characteristics which have been lost to technology's methodology and causes opposition to what else might be lost, thereby giving rise to fragmentation. Technology intensifies the differences between citizenship based upon blood, that is, citizenship determined by the citizenship of one's parents (*ius sanguinis*), which calls into doubt the legitimacy of immigration, and citizenship based upon the place of birth (*ius soli*), which challenges and dilutes the idea of ethnic identity. Although minorities in a polyglot civilization become particularist, demanding separation rather than assimilation, their very presence makes the majority question their own sense of identity. Technology throws strangers together who are forced to occupy the same space, but this proximity does not make for a symbiotic or harmonious relationship. Although man has always been a traveler and colonizer, technology challenges and modifies such tendencies. Not only are we uncertain of each other's status, but we also are distrustful of one another. Beneath this distrust lies xenophobia, an intense fear of foreigners, and fear, a most powerful force at our disposal, is often used to solidify one people against another.

Since progress is perceived quantitatively, it is considered as a means to having more, not being more, and having more is considered better because it allegedly brings an increase in power. Since being becomes secondary to having, the triumph of having at the expense of being is to our detriment. Each advance in technological progress has been facilitated because being has been forsaken, because we no longer are preoccupied with it. Technology and the technological age are well suited for the development of things, enabling us to perceive possessions as refections of being. As Sartre has indicated, we think we are what we have,[4] but such a notion is a grave distortion because having can never be more than an extension of being. Under this misconception, we have the idea that consumerism is a most useful benefit to us. As technology's handmaiden, consumerism is presumed to encourage the growth of being because more things are brought into the world by its agency. Of course, the opposite is really the case, since technology and consumerism suppress being by denying its rightful place in the world.

More to the point, technology has created a crisis of value; it is the cause of our metaphysical disorientation. To put it simply, we have forgotten how to say no. Because technology is compulsive, we feel driven to do whatever is possible. We feel compelled to be the vehicle of technology's cogent power, by pursuing a predetermined manner of operation, by adhering to a specific mode of activity. And technology's power is as mindless as it is irresistible. So we have become the objects of domination, the objects of a demanding institution, which when formed, manufactured and marketed, makes us subordinate. Technology's raison d'être assumes that if we can do something, we ought to do it. And if we ought to do it, then we must do it. It is for this reason that technology limits human choices--for if we are powerless to resist technology's latent power, we can hardly call ourselves free.

Notes

1. Descartes, *Les Principes de la Philosophie*, pt. IV, art. 203, *Oeuvres philosophiques*, ed. Ferdinand Alquié (Paris, 1973), 3, p. 520.

2. Bacon, *The Great Instauration*, preface in *The New Organon and Related Writings*, ed. Fulton H. Anderson (Indianapolis, 1960), pp. 7-16.

3. Heidegger, *The Question Concerning Technology*, trans. Lovitt, p. 21.

4. Sartre, *Being and Nothingness. An Essay on Phenomenological Ontology*, trans. Hazel E. Barnes (New York, 1956), p. 591.

Chapter 3

Termini Ad Quem: The Limits of Technology

Technology is accompanied by the belief that anything is possible, that we can perform all tasks, make all manner of machines, master everything. And because so much has already been produced by technology, we also think that everything conceivable is possible, that we are in no way conditioned by circumstance, in no way bound to limitation. In short, our age is characterized by intellectual arrogance which contends that there are no limits to technology, likewise, no limits to progress. We are convinced that technology can do just about anything, that it allows us to manage all roles and solve all problems, that we have entered a golden age with ourselves as its masters. Thus, an illusion of our age is that if something technologically can be done, it ought to be done and indeed must be done. But what we can do should not be equated with what we must do for in no way are we compelled by unfulfilled possibilities. That is of course an illusion: an action is neither restrained nor exhorted by the existence of a possibility because a possibility is not an imperative. Moreover, in addition to our illusions, we are prisoners of certain real limitations from which we cannot escape. There is no technology that can alter the circumstances of time past in which we were born, nor is there any way of preventing the ultimate possibility of our death. But if we fulfill only a technological purpose, then we have value only when we are useful to technology. Possibilities are defined by certain conditions, only one of which is choice, and they merely tell us what might happen, not what must happen. No choice is limitless, for even history is made within a range of possibilities. Just as we can speak of law and the limits of freedom, to cite an example, we may also speak of circumstance and the limits of technology.

Although technology is quite capable of fulfilling material necessities, it fails to fulfill non-material needs, to uplift us to a higher plain, to overcome our inherent lack of completeness. Its uses, its very design, are practical, and it is to practicality that it addresses itself. Technology is built upon the premise that whatever has no practical application is worthless to technology's pragmatic mentality. But if we recognize that there are limits to technology's application, we should also recognize that technology merely constitutes a basis for itself, and by no means a basis for us. Therefore, our ultimate hope cannot lie with technology. We cannot be conditioned by a phenomenon whose sole purpose is the perpetuation of means with no end in sight.

Because we feel compelled to select from among only those possibilities made allowable by technology, we lessen our freedom of choice. As technologically primitive as our predecessors were, they had greater freedom of choice insofar that they were less hemmed in by technology. It is true that they could not split the atom or walk on the moon, but it is also true that they were not defined merely by technique, not bound to a specific way of doing everything. Although it is true that man is always accompanied by some form of technology, we do not have to be captivated by it. We ask ourselves why we did not think of a particular technology or a technological approach before we had in fact devised it because it seems after the fact to be the normal way of doing things. After we devise new technologies, we regard them to be very much an integrated part of what we know and experience so that we find life to be inconceivable without them. This is the way, for example, we perceive the invention of the electric light bulb when compared with the wax candle.

Although our predecessors could not do as much technologically because they had simpler machines and fewer techniques, they could act in ways that we do not--that is, they acted more spontaneously, more haphazardly, less intrusively. But technology always limits choice because it makes only certain options available. Not only that, but technology also suggests what choices it favors, that is, the choices that we should make. And thus it is we who put unrelenting pressure upon the fulfillment of only those possibilities open to technology, causing us to act within a smaller and smaller circle of possibilities the more we are consumed by technique. Because the use of technology gives us the impression of being unlimited, open and unconditional, it is a source of the blatant optimism of our time. But the reality is the opposite: technology acts within a limited and closed system with decreasing alternatives and diminishing returns. We have yet to realize that technology cannot set us free from our pains.

The more technology becomes efficient, that is, the more technology becomes technological, the more it becomes vulnerable. Every increase in technology means an increased dependence, which signals a heightened vulnerability, but the more we are tied to technology, the more we defend it. We put ourselves on the defensive whenever it is questioned. Ironically, our defensiveness makes technology open to attack, liable to injury by less technological means.

Technology can only exist within the limitations of its possibilities; it is not limitless. Nature itself places limits upon human activity because its resources are not without end. In a world of finite resources, our choices cannot be infinite. We live in a world with limited, unrenewable resources, such as coal, oil and gas, and if we are careless, the land as well. Because our use of energy is linear, we cannot return natural resources from where they came after we consume them. Under these circumstances, energy consumed is energy destroyed--in effect if not in fact. Our exploitation of nature always produces scarcity because the more we take out of the earth, the less we leave. Contrary to Marx, scarcity cannot be overcome by man's mastery of nature, although we can agree with him that scarcity is a cause of competition. Scarcity results because of the depletion of natural resources, making goods more expensive and more tied to economics.

Since a technological age ties all regions of the world together economically, it links all nations into an interdependent system so that raw materials and finished products, from grain and oil to automobiles and computers, become global commodities. A deficit in one economy or the worst calamity--economic collapse--would increase pressure in other economies to offset it. Not surprisingly, self-sufficiency is virtually impossible. Scarcity caused by our use of natural resources actively and directly links ecology with economy. For example, deforestation affects the cost of lumber and therefore the building of houses. Likewise, a decrease in our ability to produce food because of low rainfall or soil depletion affects agricultural production and therefore the price of food. The deterioration of any environmental system as the result of mistreatment, pollution or overgrowth has economic consequences. Since the resources of the earth are limited and subject to exploitation, their consumption predicts the end of our age, that is, vulnerability is predicated in scarcity. Nevertheless, the limits of technology do not eliminate the feeling of domination, which so very much characterizes technology when we think of it.

Regrettably, technology is not considered to be limited by size, yet size itself may be one of its greatest hindrances. Beyond a certain limit, size becomes an obstacle. The burden of size increases faster than those resources needed to keep pace with it. But as fast as technology moves, it is followed just as fast by decay. Each new advance is followed by the obsolescence of what preceded it, for the latest advancements tend to replace all former ones, just as the automobile has replaced the horse and buggy. Like Cronus, technology is the parent who devours his own children. It brings things into the world only to rid the world of them. Hence, technology produces little of lasting value. It is for this reason that technology is always tied to economics, tied to the needs of production. Yet production is meaningless without consumption. Goods and services must be both bought and sold, made and consumed. Since production and consumption are inherently connected with technology, a technological age is designed for continual production. This is why mass production and mass consumption are phenomena of our time. Goods and services are brought into the world so that they may be consumed, very much like the energy which produced them. And what is acquired is soon discarded because it is not made to last, because it has lost its marketability, because in the very act of being acquired it has been devalued. Like technology's definition of history and its view of progress, the concept of production is also linear, and linearity leads to waste.

To be potentially all things by means of a technology that is not tied to limits is to be potentially nothing because limits help to delineate possibilities, and choice determines which way we will go down the road of time. Technology is applicable only within a temporal and finite world, but we utilize technology as if it has no limits. Since limitations are everywhere, technology must be subject to the same or similar conditions regardless of its impression to the contrary--which include such things as natural resources, time and use. To claim otherwise is ridiculous. It is equally ludicrous to hope that technology can be all things to everyone, the solution to all problems, or omnipotent in all situations, especially when it beckons with the invitation that we surrender ourselves to it unthinkingly.

To say that technology has become more vulnerable with every perfection is to say that technology has made the world unsafe. It has removed one type of danger only to substitute another. It has removed one type of uncertainty to replace it with another. But more than all of these, it has oppressed the manifestation of human beings; it puts up barriers to existence and restricts human choices. Because technology is not conditioned by some distant idol, unapproachable and incomprehensible, but by ourselves, the restriction of choices becomes more binding upon us so that the circle of choices becomes smaller with every advance in technology. There is no contradiction in saying that our choices are restricted by technology even though the world is expanding. Intellectually, the world, like the universe of which it forms a part, increases in size because of human effort and time, but simultaneously decreases because technology imposes limits upon those choices immediately open to us. We might see that there is more out there, but there is less that we can do to assimilate what is available. Since we all have to choose from limited options, the choices we do make negate to an increasing extent what we do not choose.

Technology creates the illusion that our age is the greatest age that has befallen man, that our lot is getting better, that our place in the world is always improving. We may call this phenomenon the illusion of progress, but what some people see as progress others see as madness. Indeed, we live in a time unlike any other. We live in a time governed by mechanization, artificiality and regimentation, which in large part causes the exhaustion of humanism, the dehumanization of art and the impotence of man. It also confers a sense of irresponsibility, whose presence denotes a lack of freedom. Disillusioned, we have become careless and carefree, riding the whims of indifference while believing that technology will always protect us and keep us free from harm. We also live in an illusion of sophistication wherein friend and foe appear alike under the thin veil of civilization.

We all know that a simple or primitive technology poses fewer problems than a more advanced one because an advanced technology makes the world both convoluted and complex. A simpler technology may be used to improve well-being without causing appreciable harm to itself or its environment. But a sophisticated technology creates problems technically difficult to solve. When faced with a problem, a sophisticated technology proposes solutions only through involved technological means because problems have become increasingly intricate, that is, technology proposes solutions only within its limited sphere of influence. Consequently, it is pushed to its limits to solve problems for it creates an imbalance if it fails to do so. Because the impact of technology has increased quantitatively to such unimaginable levels, this impact attests either to the sophistication of our know-how or to our arrogance. Technology has an impact far beyond its immediate end, an impact in excess of its time and place. And by this we mean that one of the most important effects of technology is repairing its misuse in former times, in a sense to make an attempt to erase its footprints in the sand. So we import exotic or nonindigenous plants to hold the earth eroded through human folly, and when those plants invade an environment we did not intend for their presence, we import other exotic beings to hold them in check. And so on and so forth until we have caused an imbalance in nature far greater than the simple erosion of a few hills.

There can be no mistaking that the use of technology is the source of pollution so evident in our time. Pollution, like technology itself, leaves nothing alone, and intrudes everywhere. For not only can the environment be polluted, but language and morality as well, since language can be distended by bloated and empty words, or elongated by a certain type of legalese, and morality, like advertising, can be pumped up through deceit and meaningless rhetoric, enlarged through fabrication, and extended by illusion. But pollution is most formidable in nature. It endangers the earth because it forms no part of the ecological cycle between life and death. It is the product of a linear progression from a natural beginning to a toxic end. It is the symbol of the collapse of nature. We permit it because we believe we have no affinity with nature, because we regard nature to be alien to our own well-being; we use nature, but we do not protect it. We become the source of self-generated or at least self-aggravated environmental problems.

Technology helps us to strip nature of her naturalness. Since nature does not perform any technological functions, despite its multitude of natural processes from the purification of water and air to rendering habitats for plants and animals, it must be opposed to technology and progress, and therefore antagonistic to man. We exploit and abuse nature because from a technological point of view it stands in the way of our development. Nature is not to be obeyed, but conquered. It is merely an obstacle to be overcome, as when we slash our way through a jungle to build a road. It is true that nature was hostile to man before the advent of exploitation, yet we still experienced some sense of unity with it. But now exploitation has become possible because we have somehow severed our ties with nature.

But more can be said about our relationship with nature. We are ignorant of nature because we live too far from it. So we are indifferent to it. But we are also callous. We mourn very little or not at all for habitats and wildlife when lost. They are both gone and forgotten. One cannot avoid the conclusion that we see nature as the other, particularly when we want to control it. Even environmentalism is not really concerned about nature in itself, but about its impact upon man, for the problem of pollution is considered to be more important than trees, just as toxic waste is thought to be more compelling than wildlife. Our extensive scientific knowledge of the interdependence of all living things has not hindered us in the least in the exploitation of nature. Our knowledge has not given us a humility when dealing with nature, for even national parks and wildlife sanctuaries have been established deliberately and purposefully as a result of our technological view of the natural world. Wilderness as we now know it is isolated, artificially set apart from the rest of the world. We try to save what is left in a most technical manner. We restock lakes of fish, airlift animals from here to there, and engage in artificial insemination to assure offspring, and we are enabled to do all things by means of the very technology that altered nature from its natural state. Blinded by our power to cause great and lasting changes to nature, we can no longer see that nature in all its magnificence when devoid of human influence possesses a type of timelessness, like a segment of time without temporality, as if it were a piece of eternity in a temporal world.

Although technology purports to improve upon nature, the forces of nature do not wait simply for our arrival in order that they may be developed by us. Nature is not merely a resource silently awaiting exploitation. Although technology may be a mode of discovery, nature certainly is not. Trees are not there solely to be cut down and made into paper, oil and gas are not there to be consumed as a source of energy, minerals are not there to be mined and made into malleable objects, water is not there to flush away the pollutants of industrialization. If nature were a resource patiently awaiting exploitation, then nature would serve only one master--man. To all other creatures, their appearance on the earth would merely be incidental.

Without a doubt, modern technology has greatly altered our view of the physical world. Nature has been objectified to such an extent that it is now exploitable, since it is quantifiable and scientifically measurable. The final proof of such a view is that the objectification of nature promotes the belief of our mastery over it. Nature has instead become a resource for our benefit, a storehouse of potential energy. Consequently, nature is no longer thought to have a purpose in itself, but awaits the purpose which we will give it. Nature is seen simply as an asset to be developed, as a benefit for our advancement. Technology is the antecedent to which nature is the consequent. Such a view is consonant with the biblical view of man that we are the culmination of nature, that everything in the world exists solely for us, that nature has an inherent design which would be contradictory and even impossible without us. This exploitation and manipulation of nature, inherent in the underlying forward thrust of Western technology, is attributable more to the presence and influence of Christianity (and its roots in Judaism) than to any other factor. The book of Genesis (1:28) encourages man to be fruitful and multiply, to become masters over all living things, and to subdue and dominate the earth. Although nature was exploited before the appearance of Christianity, the latter has made its followers arrogant and possessive, which merely accelerates nature's exploitation. Moreover, Christianity's influence was augmented by seventeenth century Cartesianism and eighteenth century industrialism.

Despite its irresistible forces, nature left in a natural state is always in equilibrium, since one manifestation is always posed against all others, such as population growth in relationship to food supply. The occurrence of this equilibrium is true even in the aftermath of environmental degradation in which nature attempts to correct adverse effects. But the presence of man has endangered this equilibrium even more. Habitats do not have to be damaged beyond repair before they lose their ability to support life; subtle changes are often sufficient to indicate serious loss of diversity. Of course, extreme examples of environmental disruption and climatic changes have more devastating effects by both endangering the earth and threatening life. Hence, nature's equilibrium has meaning only when it has a chance of recovery, that is, when nature can be itself. The equilibrium of nature was disrupted less by the technologies of former ages. Modern technology, on the other hand, knows no moderation; it is governed by no system of self-regulation. It simply forges ahead, proceeding mechanically, almost blindly. Rather than letting nature be itself, modern technology modifies it, exploits it, abuses it. It conceives redirection as its only possibility. Therefore, technology symbolizes the distortion of nature. It is incapable of self-regulation because it progresses linearly, and nature, as we all know, is cyclic. In the true sense of the word, there is no such thing as waste in nature because in nature, when free from pollutants and toxins, everything is used by something else. Nature in a phoenixlike fashion regenerates itself; it shows that life always follows death.

Although environmental damage has occurred in all former ages, even in those times far less advanced technologically than our own, former ages appear to have had a greater respect for nature, even those ages influenced by the linearity of Christianity. Our age is like no other. Its exploitation puts it in a class by itself. Here is a principal difference between our age and all former ages, for even environmentalism is far older than conservation. Previous ages were concerned primarily with acquiring knowledge of the workings of nature in order to live as well as possible, and only secondarily at best with obtaining power over nature. Our age, on the other hand, is concerned primarily with power, and only secondarily with knowledge. Disrupting the balance of nature means nothing to us as long as we can control it.

But technology is also the means of avenging ourselves against nature, for nature's exploitation manifests man's internal crisis. It shows that we are alienated from the natural environment, in part, because we are alienated from ourselves. And because we can be abusive to each other, being abusive to nature comes without much difficulty. If we can be aggressive with others, certainly we can be aggressive with nature. Technology's exploitation of nature imposes a new dependence, a dependence which has led us into an unnatural way of life. It disintegrates our relationship with nature, and nature's relationship with us. The domination of nature really signifies the domination of man; technology can be seen as a circumventing route to enslave us by enslaving nature. There can be no separation between the exploitation of nature and the exploitation of ourselves, since how we act with nature is merely one manifestation of how we act with each other. And whatever technological achievements we make in our control of nature are accomplished at the expense of any self-control--that is, our conquest of nature occurs with the appreciable absence of any inner conquest.

Because the earth's ecological systems took millions of years to evolve, all forms of life are tied together in some way, and nature forms a single, integrated whole. All living things, as we know them, belong to the earth, not the reverse. Yet because the interrelatedness of plants and animals is subtle, this subtlety is often overlooked. Certainly no species can be separated from others, which also form part of its habitat, and most definitely no species can be separated from its past. Hence, the extirpation of a plant or animal species may lead to the extinction of others, since extinction links the fates of many different species together when those species share a common habitat. Furthermore, the alteration of habitats through geographical redistribution or transplantation of species is now happening on a global scale. Since the success of a plant or animal is often measured by the extent of its geographical distribution or the diversity of its habitats, it would be true, but absurd to say, that our destructiveness had generated successes in biological experimentation. Perhaps it is more to the point to say that our long-time extinction of other life forms is a rebuke of our ruthlessness. The mindless brutality of nature is surpassed by the deliberate brutality of man. Of course, it is an unanswered question how much change the earth can tolerate, since we do not know the long-term effects of environmental changes caused by technology. In fact, we cannot learn what the long-term consequences of our actions are to nature unless by comparison we learn what we are doing if we do nothing. And the latter, we hasten to add, is totally antagonistic to technology's premise, which is that if we can act, we must.

Although man has very adaptive qualities, the effects of sudden changes brought to bear upon other species have been devastating. Those species, notably, cats, dogs and rats, which have adapted to a man-modified habitat, will always be more successful than endemic species never or rarely influenced by man. Species that have become adjusted to living in close association with man have become just as adaptive as he. Some of them have even become pests, but, of course, the worst pest of all is man, who invades every habitat and threatens every ecological niche. Ironically, the nineteenth century, which gave us much understanding of the evolution of life, also was the same century that greatly accelerated the demise of the earth. As a result, the future will be biologically less rich and less stable. It will be no exaggeration to say that in the future the only things remaining in the world will be man, his pets and livestock destined for slaughter. All other forms of life would have passed into extinction, or be tottering helplessly on the brink of no return. Because diversity is the raw material for new life forms, life in a technological age is tediously uniform and monotonous.

Man's biological success intensifies the threat of technology. Our evolution has been so advantageous that we have managed to greatly reduce wilderness all around us. We have put so much pressure upon other life forms as to run them right out of existence. The dodo, passenger pigeon and sea mink yielded to this pressure, for once a species is endangered, it has little chance of recovery. Even the great herds of the past, thought to be beyond number, such as caribou or bison, have been reduced to a fraction of their former majesty. Even those few species brought back from the brink of extinction are saved solely through technological means. We cringe at the thought that technology would stand by and not lend a hand in their salvation. Unfortunately, nature in all its abundance and man with his technology cannot coexist because technology leads to the exhaustion of the earth, to its absolute utilization and subjugation. Thus, the antithesis of life on earth, which is both diversified and prolific, is technology. Because technology is consumed totally with means to the exclusion of any end, the final and absolute end of technology, the end beyond which there is no other end is death. The death that technology brings--be it to the natural world, to the diversity of human culture or to the multiplicity of life--is apart and beyond any occasional benefit it may confer upon us, such as the benefits of medicine and agriculture. The endangerment of all living things, the destruction of wilderness and the creation of wasteland everywhere are all the result of technology: they are all manifestations of the desecration of the earth because technology is the means of ecological destruction.

It is true that we have caused the extinction of many species, but the human species, ironically, is more mutable and less stable than those species now extinct. This is true because although in a healthy world animals lack nothing to aid their survival, we are forced to make ourselves every step of the way, from the cradle to the grave, every second of our lives. We are given the possibility of existence, but never the actuality of it, for the actuality is what we create ourselves individually. And this means, in addition to the use of freedom, in addition to the existential requirements of human reality, in addition to making a way for ourselves in the world, technology requires the use of the earth made manifest by individuals and societies. We are just as free to use the earth wisely as we are free to use it carelessly. We are just as free to live within the limits imposed by the earth or our institutions as we are free to ignore them. We are just as free to say yes as we are free to say no.

Nature is neither immortal nor impervious to our presence. If nature represents the best meaning of the wisest man as Emerson described,[1] then we live in an age of fools because the more we strike blows at nature, the more we hack away at our own defenses. There is a direct and immediate connection between our health and the health of the environment. The closer we are to nature, the closer we are to ourselves. Since our strength comes from nature, like Antaeus who remained strong as long as he had contact with the earth, distancing ourselves from nature only weakens us. It goes without saying that the only animal who arrogantly attempts to interfere with nature while simultaneously remaining grossly ignorant of the consequences of these attempts is man. It follows, therefore, that there are a number of fallacies associated with our conception of nature.

One of these is the frequent assumption that nature is to blame for our own failures, especially when a major disruption occurs in nature that leads to human misery, such as famine caused by crop failure, erosion or desertification. When these tragedies occur, we accuse nature of abandoning us, but in fact it is we who have abandoned nature. Another fallacy is that we talk of the encroachment of wilderness into the nonnatural world, but in fact it is we who encroach. Above all, we have failed to understand nature. Our failure to solve problems caused by technology indicates that our solutions are devised only to address individual aspects of a complex whole. When we consider nature, we do not think of its totality, but merely of its individual parts. And this is one reason why we have failed to understand nature. Our impact on nature shows that we act without being aware of the consequences of our actions. Because we have created our own environment through the aid of technology, totally cut off and alien to nature, we believe that we do not need nature for survival. But the products of our artificial environment are derived from the natural environment, be they in the form of fuel or water or air. We merely show our ignorance when we indicate that it is control of nature that largely occupies us. When we boastfully propose to control nature, we merely give credence to the illusion of power. But control itself is the wrong word when dealing with nature, since we should strive to achieve harmony. What we have attained is precisely the opposite. Control is simply another name for contempt, and contempt breeds violence.

Characteristically, in this technological age we do not feel restrained by a deference toward nature because in point of fact we do not feel bound to the earth; technology is not conditioned to a Stoic this-worldliness. We do not yearn to seek a harmonious relationship with nature because we have devalued it through our careless disregard of anything natural. We do not wish to be purified. We do not wish to stand in awe before the forces of life and death. Nevertheless, nature untouched represents the way the world was in the forgotten past before the advent of man. Of course, there can be no segment of time when the world stood still, since nature is never timeless. But we are latecomers in the evolutionary process, recent additions to the evolution of mammals and primates; above all, it is apparent from the way we use technology that we have been deprived of the experience of evolving from the void.

Despite an age which so easily surrenders itself to technology, we still crave pristine places, unaltered, uncontaminated, pure, where even we are visitors. We admire wilderness because it means wholeness. We know that nature symbolizes life, but we also know that we have become careless of our natural inheritance, and are at a loss how to become worthy of our sovereignty over it. Wilderness can help us realize that when we are in the presence of nature we should leave no evidence of our passing. Of course, we can never know everything nature has to offer, least of all experience it, unless we leave certain regions of the earth totally alone, unless there are places which show no evidence of our presence, as if we did not exist. If we want to preserve a place, we must make sure that there is no access into it. No machines, no techniques, no roads, no pollution must be introduced into these wild places, which give us a sense of balance because we can never know what we have created unless, by comparison, we know what we have not created. But even in the most hospitable landscapes, even in paradise, we are merely strangers. We use the earth for a very brief time and then pass beyond it, for all of us are always in transition to somewhere else.

Notes

1. Emerson, "Method of Nature," in *Complete Works*, ed. E. W. Emerson (Boston, 1884), 1, p. 204.

Chapter 4

The Point of No Return: Progress and the Linear View of History

Technology is invariably linked to the idea of progress, that is, to the idea of change; in fact, the need for change is often believed to be a cause of technology. That need may signify some type of lack, but need itself does not necessarily preclude accident or circumstance, since coincidence or any condition concomitant to another does not eliminate change. The fact that both accident and circumstance are unscheduled may not eliminate the need for change when necessitated by condition, as when, thrown into an uncomfortable situation owing to no fault of our own, we do our very best to flee. Accidents and circumstances are temporally relative and factually conditional. When associated with technology, however, change is deliberately sought and artificially created, and we call this intellectual innovation. Change results because we have both the use of reason and the ability to plan for the future, but we often step out of our bounds when confronted with change, too easily forgetting that excessive change leads to sterility, although ultimately all changes hold us captive. We must make changes only when they are necessary, not for the sake of change. To be meaningful, change must be mitigated with moderation, that is to say, change and its affiliated idea of progress must not become commands. They must not so compel us as to afford us few avenues of escape. Since our need for change can become all-consuming, we become victims of change when the latter becomes a technological imperative because the need for change interferes with our ability to say no to further changes.

Although created by progress, change is itself accelerated by that type of change which is naturally evident in a temporal world, since progress merely supplements those changes which result from a world always in a state of flux. As is always the case, a world affected by technological change does not preclude the influence of natural change. Technology cannot so modify the world as to bring nature to a halt. That man can permanently affect nature is evident, but the forces of weather and their effect upon all living things will continue even after the atmosphere or the oceans have been polluted. The movement of the tides has nothing to do with man. Hence, change is twofold: it is initiated artificially by us, as well as caused naturally by a world never at rest. Artificially caused change, that is, change induced by man, is often linked to the idea of progress. In fact, it is categorized as being synonymous with it. We assume we are making progress simply because the present looks so different from the past. But just because there is change does not mean that conditions are getting better, that is, that things are progressing. Change can be for the better or the worse. In fact, we can never know how progress makes things better or worse before the fact because we can only analyze its influence after the fact. And because we cannot know what this influence might be is reason enough why we should look for examples in the past.

By definition, progress means to move forward, to advance, to achieve improvement through a series of steps or gradations, to attain betterment through desired ends. Hence, the doctrine of progress infers that we are forever advancing to a higher level, to a more perfect state in relation to all those that have gone before. Each new age is presumed to be an advancement over all previous ages. Understandably, the notion of progress always implies a value judgment, that what is to come must be better than that which has passed away. Of course, there are different notions of progress: material, political or social are but a few, although material progress is closer to technological progress because political and social progress are more akin to morality. Progress is relative, since what one age regards to be progressive another age considers to be regressive. A value is given for what is desirable and a disvalue for what is undesirable. In a sense, the views of progress manifested by a particular age exemplify what that age believes is worth striving for. Although the past indicates what bygone ages thought of progress, no age can say what is better; it can only say what is better for itself. Just as the future cannot speak for us, we cannot speak for the past.

Inherent within technology's rationality is the paradox that progress will improve us and will change us for the better, regardless of the means used. Progress implies that all future actions, even history itself, are determined, that they are predictable. This view explains why we speak of the inevitability of progress, but we are not expected to question it. Like the infallibility of the pope, it is fashioned in such a way as to be beyond doubt. We are expected to accept it at face value. We see progress as a divinely driven force, an inexorable phenomenon, inflexible and captivating. But if we are honest with ourselves, we should realize that there can be no such thing as a theory of progress that will continue indefinitely. The idea of progress has a distinct beginning in modern history and will most certainly have a distinct end. Like fossil fuels, our belief in progress, which we have long taken for reality, will be exhausted. Progress, like everything else, is subject to unforeseeable changes, including the changes brought on by scarcity of natural resources and the disintegration of conditions.

Those of us who marvel at the advancement of progress praise only what we have gained, but are blind or indifferent to what we have lost. Yet a fair appraisal of our age affirms that to accept it is always to be ashamed of it, for it is no more than tolerable. Indeed, we pay a very high price for progress, because although it may free us from certain types of labor on the one hand, it ties us to dehumanizing and wasteful activities on the other. The mythology surrounding progress, that is, the convenience seemingly offered by technological development, infers that technology is the means by which we will achieve an earthly paradise, if we would only surrender ourselves to it, but the telesis of progress is not as well planned as we would like to believe. Thus, Leibniz's expression, popularized and ridiculed by Voltaire--that we live in the best of all possible worlds--signifies that everything fits into a world designed with overall perfection, that everything must fit, everything must have a purpose, everything must be for the best. But a world already perfect has no need for technology and little need for man. Nothing could be farther from the truth than this claim; rather, technological progress needs our imperfections in order to achieve perfection. Improvements result because we make mistakes and hopefully learn from them.

One implication of progress is the attempt to create one indivisible world, a world politically united by internationalism. Such a world is believed to redress inequalities, encouraging a solidarity of peoples that would be the inevitable result of the advancement of technology. But what really occurs is the opposite. Technology denies internationalism and promotes regionalism. It cannot make all nations equal, although the myth remains. There will be no universal nobleness for mankind, no uniformity that creates a brotherhood of man. Nor does technology help in the development of nationalism, unless perchance the overtness of technology is forced to submit to isolationism, typical, for example, of America before World War I or China some thirty years after the introduction of Maoism. On the contrary, it has widened the gap between the technologically advanced nations on the one hand and all remaining nations on the other. If left undisturbed, a gap between these nations will forever widen, regardless of the advances of less developed nations. Ultimately, technology augments the differences among nations, particularly, if they have wide disparities culturally. Furthermore, modern technology is not for everyone, nor should it be. There are parts of the world that are struggling to retain their ancient character, and that serve to remind all of us where the world has been.

Despite the hopes and confidence of many, progress remains ambivalent. It makes life fragmented. It shatters and rips apart the community of men. The advantages that it confers on the one hand (improved nutrition, higher standard of living, universal education), it takes away with the other (overpopulation, pollution, monolithic politics). Above all, progress begets the diseases from which it suffers, for it is accompanied by unprecedented social problems of overcrowding, unemployment, disease, squalor, crime, vice and disenchantment. The laws of progress are both expansive and restrictive, that is, they are in a dialectical or paradoxical relationship because the same laws that produce growth also produce disintegration. Progress presents new opportunities but also new limitations, new situations but new restrictions, new aspirations but new reckonings. Future progress is not a requirement because it is not in itself a necessity.

Belief in progress, which is expressed as excessive trust or blind faith, demonstrates not independence but unquestioning dependence. Belief in progress indicates how totally dependent we are on making improvements for the benefit of improvement, of making changes merely for the sake of change. Progress has become a longing, an unquenchable thirst, an itch that does not stop. Because we idealize growth, we see progress as growth for the purpose of future growth, for the purpose of consuming more, having more, producing more. Never are we satisfied. Not that we should be satisfied with who we are: we can always get better. No; rather, we are not satisfied with external things, with what we possess and what we do.

We are also not satisfied by where we go. Progress promotes the illusion of betterment through increased mobility, and ours is an age always in motion, an age never intended to be at rest. Thus motion acquires a value in itself. But an age seemingly on the move has little time for reflection, making it hard to think things out, particularly when we need to reflect upon our own risks. It is only when at rest that we have the optimum opportunity to think. In fact, what mobility demonstrates is that an age always in motion makes little substantive progress. Despite high speed travel, we are an age going nowhere fast. All the changes we have experienced in our time have altered the world very little, since a technological age is largely rigid. We have arrived at a plateau of relative stability, raised up above but indistinguishable from its surroundings. We have created a new order that is bound by old constraints. Although situations are presumed to be changeable simply by the nature of their being, a technological age to a large extent remains unchangeable.

No irony is meant by saying that a technological age fosters change so long as things remain the same. Although we often hear that our age has undergone much change, this claim is true only initially. Once these initial changes have been accomplished, there is little necessity to institute more change because substantive changes, that is, real not apparent changes, threaten technology's stability. The use of technologies such as telecommunications, motor travel or electricity are only slightly different from the use of those technologies in the past. Although we have now more options with telecommunications, more varied motor vehicles and more gadgets which use electricity than in the past, all of these increased possibilities are direct results of the original inventions, and therefore, are expressed within a limited matrix. Although we may modify and perfect these inventions ad infinitum, they remain basically what they have been, even if their form or their appearance varies from year to year.

The modern view of progress as an ever increasing and ever expanding materialism signifies an increase in quantity at the loss of quality, the perfection of having at the expense of being. The constant acquisition of things, that is, the ever increasing pursuit of abundance by producing more, building more, having more is what is traditionally known as progress, since the myth of progress presupposes that abundance will make us happy, especially if we cease to reflect about anything, that is, if we remain empty-headed. Nevertheless, the benefits of progress do little to soothe our disquiet. They give us temporary relief, but no permanent remedy. Our age's increasing artificiality and concomitant dehumanization occur despite progress, which after all has not made us more civilized. It has not softened our manners, nor has it curtailed our competitiveness. Of course, the modern notion of progress is a child of the Enlightenment, notably from Condorcet, but unlike the contemporary interpretation of progress, the eighteenth century French notion merely concerned knowledge, not material well-being, a notion criticized and attacked by Sorel.[1] Its purpose was the growth and development of wisdom, not materialism.

Progress implies an end toward which we strive, an end allegedly better than those ends that preceded it. Hence, the pursuit of progress implies that the world--everything that is external to man--is getting better. Indeed, we do make progress, but it is technological, not ontological. Without technology, progress would be devoid of meaning, since technology is the means by which civilization conquers nature, the very method by which progress is achieved. But the sad truth is that though technology allows us every means to conquer the world, it allows us no means to conquer ourselves. Thus, we cannot avoid the conclusion that progress is unconcerned with man. It is concerned only with the world in which we live. If this is the reasoning that supports progress, than we have become its victims because it implies that the perfection of the world always takes precedence over us. Again, we are confronted with the reality about progress, the reality that organizations and implementations of all sorts are devised which are antithetical to man. But we are not saying that we should enjoy the results of progress while despising the conditions which promote it.

Apart from the tremendous advances the world has made technologically since the late nineteenth century, we have made little or no progress in understanding what constitutes the being of being human, with the exception of a handful of innovators--Nietzsche, Dilthey and Jaspers, to name a few. Although it may come as a shock to those with powers of perception that this ignorance is universally evident, nevertheless, the fact is well received for its lack of profundity and even sanctified by a so-called enlightened world. But regardless of how many luxuries are converted into necessities, there is no real progress unless it entails progress of the spirit, and this development is not technological, but existential. Because each individual must make this journey alone, with only the signposts of past travelers to guide him, technological progress does no more than put up barriers along the way. Although technology aids the standard of living, that is, well-being, it does not help us in the least to the perfection of being or the pursuit of excellence. The morality of our time is no improvement over the morality of those ages which manifested a more primitive technology, for moral progress is not necessarily associated with economic progress and certainly does not accompany technological progress.

The advancement of technology and its affiliated concept of progress have had immense influence on the modern view of history. Just as progress is linear in scope, so too history is now currently believed to depict the linearity of man in time, of a forever expanding cosmology. It has replaced the ancient idea of the cyclicity of history, a concept closely related to the natural view of the world, and which purports to renew and regenerate the future from the past.[2] The excessive idealism associated with progress as expressed by the Enlightenment (and notably by Kant)[3] is often without any real basis in fact. Because progress is oriented to the future, an age whose principal values are linked with the future is an age paying reverence to whatever tomorrow brings. The future does not mean that everything that can be will be, or that everything that will be must be; rather it signifies that everything that will be has supplanted everything else that could be. Hence, it is a form of determinism, and the more we value the future, the more we diminish other possibilities.

But we must not be afraid of the past; we must not see the past as an impediment. If we are to understand the present and bear the consequences of the future, we must be willing to face the past without prejudice or fear. In fact, the destruction of the past annihilates the hope of the future, just as Marxism defies the past by challenging it as a socially relevant phenomenon. We must be careful not to be rescued by a future that ignores the gravitational pull of the past, for whatever the future holds for us is conceivable only when gazing into the past. There is no way that we can repeal the past, no way that we can undo it. Like it or not, it is a foundation we build upon. Otherwise, the past and the discipline given over to it would be without meaning. If we are to have a history, we must have a past, which is something that even God cannot change, and only with a past can we have a destiny--but its fulfillment is not proof of the inevitable laws of determinism.

Although our notion of destiny is somewhat different from the Greek notion of the overwhelming presence of an uncalculated and incalculable absence, nevertheless, we do share the Greeks' idea of mortality because if fate is our mortality, then history is our destiny. Contrary to the technological notion of historicism, what we mean by destiny is a purely historical phenomenon because man is a temporal being. Of course, history is fateful only after it is made, and it is made by means of freedom. The fulfillment of destiny is the making of history, in which the future is given actuality and is handed over to the past. Although it is our destiny, our duty if you will, to make history meaningful, reliance upon technology as the sole means to history is a threat to history on the grand scale.

A technological age, being rigidly methodical, is not forgiving of the past, although the past is more forgiving of itself. Our age looks upon the past in a distorted and most unhistorical manner, as a barbaric time of humanity wallowing in the mud, ridden with disease, biased by superstition and fear. Only an age so inflated with its own superiority, contemptuously regarding the manners and mores of the past as inferior, could form such a distorted view. Like the slave, who does not have anything great to strive for, the past is assigned no special value because the past, like the slave, is totally subjugated to the present. But the attributes of progress or its sophistication cannot be offered as proof of cultural superiority. Indeed, if every succeeding age was technologically, and therefore culturally, superior to what preceded it, then Rome, given to imitation, would have been superior to the intellectual and artistic achievements of Greece; the feuding and warlike Germanic kingdoms which developed in the early Middle Ages would have been superior to the law, government and civic pride of the Roman empire; and the austerity of the Reformation would have been humanistically and culturally superior to the vitality of the Renaissance. But these suppositions are patently untrue--apart from the fact that it is deceptive to compare one age with another--because no age has yet reflected more than a glimmer of the glories of a former age. Even mythology places the golden age in the past.

The road to history is not necessarily paved by technology, that is, technology in itself does not signify the making of history. Although there is such a thing as history of technology, technology is not the whole of history. The view that technology and the affiliated concept of progress are the reasons for history merely reflect our age's view of what is rightfully meaningful, and our obsession with technology shows how our understanding is influenced by our views of the human condition and the historical milieu-- but it is an understanding which demonstrates that the values of our age are one-sided and slanted. Although values enable anyone to adapt themselves to their age, our values ultimately fail us because we have lost control of the age we wish to dominate. Technology is a means of adaptation and a value, but it hardly qualifies as the end of man in time.

All links to the past are demolished in a technological age as a means to hasten the future, but the future is never logically predictable. Ignorance of the past guarantees that predictions of the future will be unreliable, although in any case the future bears little or no resemblance to the prophecies conferred upon it. The modern age is forward-looking into a future which it hopes to create. But in order to understand what the future holds for us, we must do the opposite; we must look to the past for guidance because only the past can tell us what is feasible. The past may be classified into the distant past, which survives for many of us only as a curiosity because it bears little or no resemblance to our view of the present, and the immediate past, which has slipped away within our living memory. But no matter how we define the past, we can never be free from it. The past is not lifeless; it lives in us. To speak of its death is to use a misnomer.

Because our age is neglectful of the past, it is oblivious to history and occupied with passing events, rather than with the quest for understanding and the critical analysis of time gone by, but above all, a process. Our views of the past consume us like a parade, since an event which now passes before us and holds our attention is replaced by the next in line. But history is built up over time by tiny contributions, like a barrier reef in which each organism is embedded in what lies beneath. History is formed like a mighty glacier, by the accumulation of past events, and like a glacier moves ever so slowly carrying with itself everything it has accumulated.

Regardless of these distinctions, the lack of historical knowledge is universal, evident today among the lofty and the lowly. It is a lack that all classes share. Historical knowledge includes not only the dates and events of historical significance, but the manners and mores of the past; not only the names of individuals, but the values that they held, that is, the underlying morality of their time. Ignoring the past as a means by which to measure possibilities, our age forges ahead with dash and speed into an uncertain future. We are an age plagued by a system of extreme rationality, believing that everything begins with today, ignorant of the events of the past or ignorant that there even was a past. We appear to be caught up in an eternal now. Hence, the past means little to us, but it is a gross misunderstanding, even a degradation, to treat history as if it were without meaning. We have rendered ourselves incapable of deciding why one age or one event is more important than another, or even why it matters at all.

Apparently, we have little time to devote to history, and ironically, little technique for knowing it. Although our age is obsessed with technique, it lacks the technique for understanding the history of our own time. This inadequacy is due, in part, to the rushing by of events, which leaves us little time to analyze them. But more importantly, too much happens of a trivial nature, and too little of a serious nature for a technological age to develop historical technique. Since a technological age is essentially static, apart from being bombarded by the trivia of the moment, it produces little meaningful change. In fact, its aim is to produce a condition whose purpose is the maintenance of a status quo. Assuredly the desire for deliberate change as a feature of technology cannot affect the status quo, because the changes brought on by technology are self-serving. Technology thinks only of itself, since it desires to create an atmosphere in which only it will thrive. Artificial, deliberately construed change perpetuates further artificial, non-essential modifications so that real, meaningful change has no chance.

And because we wish to break with the continuity of the past, we have lost the sense for the reason of history, and the sense that history means change. Since a technological age is largely static, it does not lend itself easily to the making of history. It avoids the fact that reliance upon the past is how the future comes into being. The very cause for history escapes our notice because we wish to start out anew while ignoring all the steps that led up to the present. Yet history is inescapable because it is right behind us. We cannot escape the means by which we arrived at the present. Ignorance of history implies more than just a lack of knowledge about the past. It implies both empty-headedness regarding cultural and ethnic identity and foolish disregard of those values that moved men to change the present and make the future. Ignorance of the past places one in a vacuum, cut off and estranged from the world, for the knowledge that elucidates the workings of the world, the intellectual discipline which describes, investigates and explains human phenomena is not any science, but history. And history is the result of both freedom and culture; the former is ontological, the latter moral. Thus, history is explicitly irreversible and implicitly evitable.

A fair assessment of the past gives us a different perspective. It indicates that our views of the past change every time we change our values. Likewise, our views of the future undergo similar changes. Although we often study the things of the past in terms of errors perceived in the present, we also study the past because it speaks to us anew. But at all costs, we should not read backwards into the past simply to justify a reading of the present, that is, we must avoid attributing a morality to the past antithetical to its values. Because there is peril in projecting the present upon the past, the latter must be depicted truthfully and in conformity with the facts, but the facts must be put to the test. Just because a fact exists does not mean that it should stand uncriticized. Since technology is a linear process, we both act and react to its presence. Because everything in our age is permeated by technology, we can only understand our time if we understand this permeation. Technology alters the character of social life because it has a symbolic dimension that gives it meaning, but a meaning that often conflicts with reality, most notably, with technology's alleged cosmopolitanism.

Indeed, it is comical in the extreme that an age so permeated by technology attempts to bring diversified peoples together from all over the world into a common civilization only to shatter it later. A technological age is immiscible. It leads to fission, not fusion. Its subjects are incapable of attaining homogeneity. It makes everyone ethnically and racially conscious, that is, technology makes us more aware of ourselves: it enhances a greater awareness of not only who one is but also who one is not. Although racism should never become respectable, it is a direct result of life in a technological age. In fact, there is an appreciable difference between racism in the past, which was based on ignorance, and today's racism, which is based on confrontation, upon a kind of face-to-face conflict. A technological age is truly an age of pluralism, in which different ethnic, racial and social groups maintain their identity on the one hand while participating in a common civilization on the other. The dream of a universal brotherhood in which everyone would put aside their differences and work toward a common goal on more than a superficial level is contrary to technology's particularity. The attainment of such a brotherhood is unlikely. Of course, a politically democratic multi-ethnic and multi-racial pluralistic civilization is not a victory for mankind, but a permanent obstacle to greatness because a social egalitarianism in which all people intermingle produces a monolithic culture, a massive and uniform obstacle to man's betterment. This common civilization, this democratization, is most representative of technology in the West and a cause of its sterility.

Because technology cannot create a universal brotherhood, it settles for what it can create--a civilization fashioned for the masses. Since a technological age leaves no nation in isolation, its great attraction is to bring all peoples together into some sort of disharmonious harmony, knitting them into a whole ever so superficially, linking all to the globalization of the world. This worldwide tendency, which is known as westernization, aggravates the dilution of the world's cultures, no matter how unintentional or remote the cause of this may be. So wherever we travel, we encounter relatively the same things, even if we venture to the far ends of the earth. Technology is guilty of challenging both the awareness of culture and its historical basis. And thus it becomes a threat to history on the grand scale, to that type of history characterized by the magnanimous purpose typical of a classical age. If we are to achieve greatness, to be great and to do great things, we must look within; we must internalize both ourselves and our culture. Rarely do great things have great beginnings; we must achieve greatness through a directive of single-minded purpose by mastering internality.

If the West is quickly losing its sense of cultural identity, it is because it has been betrayed by the sentiment of guilt, which is easier to bear than helplessness, since we too readily receive people from too wide a background. To imitate others of a particular culture, that is, to show the externality of another's culture, is quite different than to be a part of that culture internally. Just as there are different levels of acculturation within the same civilization, there are also different levels of assimilation. Wearing a burnoose does not make one an Arab nor does speaking Mandarin make one Chinese. Any civilization must not be confused with its acculturation, which means for us that Western civilization must not be confused with westernization. For not only is it impossible for high art and culture to appeal to the masses, it is equally impossible for a group or class of people to maintain a culture foreign to their heritage. But the greatest attraction to life in the West for those who were not born there is no longer culture or religion or ethnicity, but economics. They come to the West no longer for love of it or its ideology, but for love of its money, which hardly qualifies as the best reason for immigration.

Hence, it is our fate to be situated in a world that has no need for purity, no need for conscious self-awareness of our blood, our obligations, our history. If we are to be loyal to ourselves, we must be loyal to the past. We must not demean it, because when we are no longer loyal to ourselves, when we accept and integrate everything, then we will have lost our identity, for those who cannot affirm their values have lost them. We need only to cite ancient Greece, or Rome before its empire, or Renaissance Italy, or nineteenth century Germany as examples of cultural purity and correspondingly high achievement. The adulteration of a people produces at best a mediocre culture, devoid of greatness, without differences or creative diversity. Purity and self-awareness go together. Thus, ethnic, religious and racial differences are not fictions, but historical facts. They are at the core of history. A study of any civilization, ancient, medieval or modern, manifests the truth of these facts, although what is often true may not be politically correct. Sadly, we too easily encourage our youth to ignore these truths, because they embarrass us, because they cut too close to the bone. A denial of these truths is a denial of history, a denial of a people's identity. Indeed, if we do not speak for ourselves, who will? And if history shows that we have done something shameful in the past, let us correct it, but if it shows that we have done something great in the past, let us repeat it.

If these conclusions seem to manifest a kind of determinism, let us admit that every human being, in addition to inborn differences, comes into the world with certain conditions already waiting for him. For surely we are all born to certain parents, to certain places, at certain times, historically and culturally existing before we do. No one enters the world as a tabula rasa, for even a historical process waits for us at our birth. All of these conditions, including our mental and physical powers, our race and sex, go to make up what we call the self. They distinguish you from me. Collectively, these conditions go to make up a people, who lend their hands to history. Since we are all navigators on the rivers of time, what we do, individually and collectively, will determine which way we will go. But at all costs, we must avoid the embankments of indifference and the harbors of guilt. We must catch the meanders of the past and steer into the future by their currents.

Notes

1. Condorcet, *Outlines of a Historical View of the Progress of the Human Mind*, 10th epoch (London, 1795), but see the criticism of Sorel, *The Illusions of Progress*, trans. John and Charlotte Stanley (Berkeley, 1969).

2. See, for example, Seneca, *Naturales Quaestiones*, III, 30.

3. Kant, *The Conflict of the Faculties*, trans. Mary J. Gregor (New York, 1979), p. 159.

Chapter 5

The Crossing of the Styx: Stability, Sterility and Death

In the not so distant past, we supposed that social conditions determined technology, but this is no longer true. We are now faced with a technology that seemingly alters itself independently, that is, society now follows technology. Although in theory alterable, in practice technology is rigid because its flexibility is manifested only within the perimeters of its rationality, because it is evident only within the boundaries of its methodology. We are actively encouraged to make changes for the sake of progress, yet the technological age which advocates these changes itself remains fixed. The substantive changes which occurred in the past before the full flowering of technology as a system are behind us; any additional changes we now make will be small by comparison. The text has been written. What remains is exegesis. Consequently, we have reached a plateau which demands that no genuine, no real meaningful changes be made. Technology's methodology is designed to prevent the world from changing, particularly from changing in any way that technology disapproves of. There is no paradox here between the changes we are expected to make and the rigidity that we ultimately must confront because technology renders powerless and useless the normal function of things. Although wasteful, technology epitomizes an institution devised to be in agreement with everything, which explains why all corners of the globe are more similar than dissimilar, and why technology is a threat to the infinite diversity of the world's cultures. Technology's preoccupation with technique, regardless of who or what is disposed to it, inevitably leads to standardization, which in turn causes conformity. It molds the world to itself after imposing itself upon the world.

Apropos of change, technology promotes the illusion that it is able to respond to changing situations, that it is able to take emergency measures in an endangered world, but in fact, technology is slow to act and slow to remedy problems, and slower still to remedy problems directly caused by it. Being slow to admit that it could make mistakes, technology falls victim to the same imperfections it finds in a world denuded of its presence. Even when it is admitted that mistakes have been made, the fault is never laid at technology's doorstep, but on some other distant circumstance, such as the poor application of technology or the ignorance of those who use it. Although the solution of a technical problem may be the cause of other problems, nevertheless, technology is always perceived to be helpful, good and wholesome. Few of us are willing to admit that technology itself is the cause of problems. We are more apt to praise it than to criticize it because the very fact that we need it weakens our critical acumen. We overlook its shortcomings because we love it so, like an affectionate parent who is oblivious to the bad manners of a naughty child. Of course, the belief that technology is not in need of criticism is illusory--we all know that new technologies are always based upon old technologies, and new technologies result, in part, because old technologies have been found to be imperfect. Yet it is considered almost blasphemous to question technology's perfection, and this lack of doubt conditions us to accept what is available through technology. Thus, we readily receive and regard as normal whatever is presented to us. Although innovation will continue in the future, it is meaningful because there already is a methodology into which it fits.

The lack of change creates a pasteurized, sterilized, homogenized view of the world in which anything threatening, anything objectionable, has been destroyed without causing major alterations to what remains. Homogenization explains why there are no great cultural changes or cultural movements in our time. Although we can still sink as low as ever, as low as world wars, as low as concentration camps, as low as the extermination of a people, we can no longer rise as high as before. Homogenization means that we can no longer produce individuals like Plato or da Vinci or Mozart; it means that we are fundamentally incapable of creating anything great. Unlike all former ages, our age presupposes that we will not dispute technology's alleged perfection, that we will not question a methodology that is devised to be beyond doubt. So we find it incomprehensible that there were times when decisive changes did occur, and that they occurred in times far less technically advanced than our own, such as that great cultural change which gave birth to the Renaissance out of the theocratic Middle Ages, or the age of exploration that opened European consciousness to the rest of the world, or the intellectual changes resulting from the Copernican revolution.

Naturally, a diversified technology possesses alternate possibilities. Although there is diversity of design even in the manufactured products of industry, these products too easily become standardized. The more a design becomes perfected, the more it approaches an ideal, a single entity. The more uniform or commonplace technology becomes, the more it suffers a decrease in alternatives. Hence, the shortcomings of a universalized technology are repeated everywhere, and this imperfection is true for everything technology touches, from its machines to its morality. Because technology dislikes differences or disorder, a technological age degenerates into a type of conformed obedience in accordance with the prevailing standard. Technology has made us acutely aware of differences on the one hand, even if for no other reason than because it attacks them; but it strives to eliminate them on the other. It attempts to destroy any differences, distinctions or dissimilarities between nations, races, classes and sexes. Everything is equalized by such secularization, even those things that are opposites. So everything, from individuals on the one hand to the world's cultures on the other, is reduced to the lowest common denominator. In consequence, we all wish to be alike: to share the same perspective, to voice the same opinions, to think the same thoughts, to desire the same destiny. Just as the natural world would be virtually dead if all species perished, leaving only one to succeed, the same fate may await man's world, the social and cultural realm in which he abides, for whatever technology touches, it strives to standardize, and this is applicable both to the world and to man.

Culturally, sterility accompanies a technological age because of inherent stability and self-assured security. Such an age excels in the perfection of methodology and rationality; it is moved more by organization than chance, influenced more by order than inspiration, determined more by requisition than predisposition--but despite these characteristics, a technological age will not lead to a single, worldwide culture because of the sheer vastness of the world and the impracticality of universalized control. Individualized expressions of national cultures will not be subjugated to a rigid monolithic type of supervision, although this standardization does signify that the underlying edifice of a technological age--no matter where it is--will be theoretically the same, that there will be an equivalent medium to which all peoples and all nations are subjected, and from which they derive an association, since technology has a leveling effect wherever it is utilized. Technology constitutes that one thing which has made itself irresistibly attractive to all peoples from diversified cultures who share no religion or history in common. Technology's omnipresent sterility looks suspiciously upon anything extraordinary which does not fit into technology's view of the way the world should look. Hence, extraordinary individuals, ideas or values, all of which are faced with diminishing opportunities of free expression in a technological age, have limited prospects of finding an audience or being appreciated. Genuine creativity, truly extraordinary intellectual powers, or values of lasting worth conflict with technology's rigid uniformity, because these manifestations are at odds with a morality designed specifically for one and all.

Technology's intensification lies with doing, not with being. The premodern view that actions emanate from being has been reversed; in the present view, being emanates from actions just as functions determine substance. And since a manner of doing is considered superior to a state of being, performance is everything. How one does something is regarded to be better than who one is. The collective elements of a thing or its manner of performance is a primary preoccupation of our time. Since techniques are variable and are subject to endless perfection as a means, nothing lasting is produced in a technological age because it does not concern itself with ends, but with the mastery of means over ends. We are an age more preoccupied with stability--or should we say, immobility--than with the free expression of creative endeavor. It would be difficult to overestimate what we really want because it is the opposite of what we advocate. We are an age that generates change so that we may remain unchanged. We are an age that continually alters the contours of things so that we may remain fixed. Like convictions, stability is unmoving, walled up as in a fortress, inattentive and blind to the changes necessitated by reality. And so we avoid these necessities and in their place, do nothing, are nothing, become nothing.

Because technology expands the world and man's place in it, it cannot help but present a fragmented view of reality. This is why the world has grown too large for any person to comprehend. In order to comprehend it, either we must forever struggle against overwhelming odds, always met by defeat, or we must return to an abridged form of the world that we can comprehend. Wherever there is confusion, we tend to grasp what is near at hand, what we can understand, what we can love, but the world's fragmentation also confounds understanding. Technology's expansion of the world removes the tangibility between men. Ironically, the sure numbers of the masses are not the only thing that is onerous to an age dominated by technology, for there is also the very inability of the world to bring the individuals in the mass together.

Our use of technology makes the world insecure because it generates the need for more security, because it is completely dependent upon the advances of progress. Yet we take little solace in the notion that technology will save us. If anything, it adds to our disquiet, since technology fosters the illusion that all we have to do to save ourselves is do nothing, that technology will improve the world and man, and will improve both without effort on anyone's part, that we may sit idly by and watch the improvements brought to bear upon us by the workings of technology's external forces. If there ever was an example of passivity, this is it, to achieve with the effortlessness of gods. Although many aspects of progress are beneficial to us, ultimately it leaves us with little to strive for because the options made available to us are truly limited. Despite the fact that a technological age fosters the illusion that we can be anything we want to be, that we can strive to be anything imaginable, many of us have no notable achievements to our credit because we have no reason to be anything in particular, above all, no reason to become anything.

Not only does a technological age lack the stability of mutual trust and understanding, it also lacks the potential to achieve it. The only stability associated with technology is that of insecurity, applicable to both man and his institutions. What helped us in the past to develop and encourage regional loyalties has been weakened by technology, since technology has a tendency to disassociate people from a feeling of community, a feeling of place, and thus it is not surprising that there is so little civic pride in our time. A feeling of community and taking pride in it are not possible for those who have no awareness of how they fit into the scheme of things. For community signifies a feeling of fellowship, a sense of membership with other people who share common interests. Thus, it is both apparent and pitiful that technology fails miserably in attempting to attain that which it longs to achieve. Because technology attacks and wrestles with all opponents that in any way are potentially threatening to it, it is left only with those allies that are akin to it-- namely, conceit, arrogance, criticism of others but never of itself. Mutual trust and understanding, the basis of camaraderie, are inconceivable and impossible because technology does not trust a dissenting voice.

In fact, technology has grown so powerful that it must by necessity challenge all pre-existing institutions, since nothing must stand in the way of this juggernaut. Institutions from former times become obstacles to the institutionalization of technology. If they cannot be improved or repaired, then they must be discarded. This is why technology has such an overwhelming influence upon everything it touches, for it either incorporates or annihilates. Technology's centralization brings regimentation to its institutions, and as the institutions become intricately centralized, risks are minimized. Social conflicts must be reduced to near non-existence. But more revealing is the fact that since these institutions are extremely rational, they cannot cope with irrational or spontaneous behavior, and therefore, are ill-equipped to handle unforeseen situations. When confronted with a situation difficult to understand or control, the response of such institutions is to bring it into line with its methodology, or failing that, to destroy it.

To receive the benefits of technology is also to be dependent upon its organization. Accepting technology's apparatus means accepting its design, which is analogous to saying that when we accept technology, we also accept the way it is organized--its morality, its way of perceiving the world, even its conceptions of time and space. Yet, it is often this very organization which hinders technology's application. All the technology in the world means nothing if its utility is hampered by cumbersome and inefficient means. So technological organization must always strive to stabilize itself. As with the construction of a building or bridge, technology proceeds with the assumption that the earth will not shift beneath it. Of course, organization is the means by which something is put into a structured whole, the manner by which it is arranged and systematized into a unity, but therein lies its vulnerability. An increase in technology means an increase in interdependence, and a breakdown in any aspect puts pressure on other related aspects. Since vulnerability is related to cohesion, the more technology ties everything together, the more dependent it becomes because mishaps anywhere threaten technology's efficiency.

Any increase in technology is followed quickly by an increase in the methods of organization that supposedly control it, and this organization itself is highly technical. Although by itself more organization does not bring more technology, more technology always brings more organization. Technology buries everything beneath a torrent of regulations. Nevertheless, the mastery of the broad concepts of organization is inconsequential to technology, since technology concerns itself primarily with the minuscule aspects of its manifold structure. This preoccupation with the more obscure aspects of organization cannot help but disengage it from the attributes which are always associated with it: efficiency, uniformity and universality. This is not to say that highly technological organization shows no concern for these attributes, but that they become both remote and alienating from any judicious basis of organization. Therefore, organization in a technological age is indifferent, distant and uncaring. Since such organization is a barrier between people and efficiency, it is a cause of estrangement.

Like all organizations, technological organization also brings bureaucracy. It strives to regulate conduct by fixed and well-defined rules that categorize everything. Since it leaves nothing to chance, it overrides spontaneity. It is regulated, limited and manipulative. And for these reasons, it is the enemy of individuality. It anticipates predetermined results by becoming a yardstick of conformity. Every aspect of a technological age, from the state to industry, from the military to the university, is typified by such organization. Its decisions, which are often contradictory, are handed down from no one in particular, and are characterized by their usual arrogance and folly. Because of their anonymity, they appear to come from nowhere. Thus, like all bureaucracies, technology's bureaucracy is characterized by anonymous, faceless men, who may not have the qualities that their responsibilities require of them, although their position may engender an exaggerated idea of those qualities they do have. And because they concentrate on their own contrivances and concerns, bureaucracies rarely admit that they make mistakes, and rarer still do they correct them.

Any organization that improves its techniques becomes more technical. Similarly, the improvement of techniques leads to more organization. Since organization implies standardization, the latter signifies that form takes precedence over content, so that its organization is really much ado about nothing. And that is the nature of technology's bureaucracy, the organization of needless obstacles on the way to floundering in idle and meaningless objectives. For not only does technology crave bureaucracy, it also leaves its mark of unending regulations and convolutions, and endless reams of paper. Although itself a cause of scarcity, technology always breeds excess, and whatever advantages it confers, it becomes increasingly costly, bureaucratic and contrived. The organization that buttresses technology makes a technological age more rigid with every perfection. Thus, its inflexibility is directly proportional to its proficiency. Once again, this relationship indicates why technology in the last analysis ushers in no truly, meaningful changes because rigidity, that is, stability, is its ideal. Wherever technology helps us to succeed mechanically and prosper technically, it fails to aid us creatively. Technology fails because it is incapable of making meaningful discoveries, because it brings no new meaningful values into the world, because it curtails freedom through the internalization of injunctions.

Techniques condition us to life in a mechanical world. So also does technological organization. And the fact that we learn to conform so well goes a long way to demonstrate that creativity is subordinate to adaptation, which brings us once again to the idea of mass culture. The implementation of technology is the manner by which individuals are mechanized into masses. Just as technology is implemented through the masses, the masses are mechanized through technology. Hence, mechanization and the masses sustain each other. They feed off each other's egos. Since any form of mechanization enforces organization, the latter in turn shrinks the autonomy of the individual. Wherever you find one process, you find the other.

So, too, technology makes cooperation more difficult because it increases competition. There is no contradiction implicit here; the increase in competition is of a peculiar type that is applicable only within the system, only in compliance with the rules of the game. Rivalry among individuals hinders common action, and the more competitive they become, the more they turn to technology in order to avoid cooperating with one another. The interplay of cooperation and competition, the interaction of these opposing forces, is like sibling rivalry within the family, because cooperation becomes more than an ideal; it becomes nearly an impossibility, unless, of course, one or all deny individual likes and dislikes, that is, free expression.

Technology's systematization symbolizes everything, but the individual represents nothing, unless he is part of the system. And since we are taught only to conform, any criticisms we have of technology must be expressed secretly, for to speak out would be a catastrophic denial, a threat and challenge to our presumed benefactor, an affront to our universal protector. All too infrequently do we question the omnipresence and omnipotence of technology. We do not see it as human, but divine, as a god which looks kindly upon us.

Because of technology's omnipresence, it comes knocking at everyone's door, invading even the abodes of the lowly. It refuses to leave any person in a natural state because it loathes naturalness. Technology compels everything to conform to it, and because conformity is one of its aims, it can demand the social character it needs. It indicates forcefully and openly what type of social character it regards to be compatible with itself--a manifestation of its dogmatism. It prescribes rules and maxims to which its members are expected to comply.

Although we hear much talk of the pursuit of individuality in our time, in fact, little or no individuality exists in such homogeneity. Indeed, how could there be individuality when everyone seeks conformity? Obviously, there are basic contradictions here between technology's presumed defense of individuality and the reality of its conformity. Yet, civil rights, so much a topic today, are advocated for everyone, for the concerned and the indifferent, for the deprived and the privileged, for the indentured freeman and the emancipated slave. And since everyone desires to be distinguished by such rights, we remain indistinguishable after we have acquired them. So we have encountered another illusion fostered by technology, the illusion that personal freedom can most assuredly increase by its means. Even if we were foolhardy to venture into such a situation, its imperfections would prevent its realization. The choices that technology offers are all within the system. Any increase in technology makes the system more, not less, restrictive. Thus, one may observe that every edifice erected for the greater glory of technology buries prospects of liberation. Indeed, the machine itself is our epitaph.

But our unquestioning pride in this edifice goes a long way to inflating our egos, increased abnormally because we too easily proclaim technology's accomplishments. Pride, nourished by an excessive egotism, is deceptive because it inhibits our understanding of what we have accomplished. It also blinds us to what else we could accomplish, since it suspends self-criticism. Because pride pampers us and gives us a false sense of protection, it makes us passive. A false feeling of security engendered by pride along with a deluded sensation of freedom go a long way to causing disillusionment. Realistically, what we should feel is the opposite because we must be on our guard against complacency. We cannot ignore the need for reappraisal. The expectations that progress can only be inevitable, that our choices are not subject to uncertainty is an illusion. Great peoples, great ideas, great achievements come into being only because of struggle, and struggle means opposition, controversy and change.[1]

Even in the present age the view of Kierkegaard is still true--that our age is a formless, superficial, excessively rational time with no depth of perception to distinguish between the obvious and the obscure.[2] It is an age in which imitation takes the place of inspiration. Because it severs all ties to our higher functions, because it curtails our ability to conceptualize, because it challenges the anticipation of needs, it is an age which lobotomizes us. If taken for what it offers at face value, the modern age is a boring place to be, where the unexpected and unforeseen no longer have meaning. Lack of spontaneity is one characteristic of our time, in which the true range of possibilities is absent. What remains is timeless repetition, relieved by occasional outbursts of violence or sensationalism. Not surprisingly, a technological age fills us with emptiness; it is characterized by the lack of imagination. It both pampers and bores us at the same time for it makes us comfortable but empty, satisfied but powerless, and above all, compliant. Most people mirror the age in which they live, and the people of our time exemplify these shortcomings.

But the mechanization of time, first evident in the medieval monastery, has itself played a part in the success of technology. Because monasticism required the observance of canonical hours in order for monks to pray and work at a fixed time each day, an accurate reckoning of time became essential. An age already accustomed to the idea of mechanization easily accepted the mechanical clock, an invention of the fourteenth century, which accelerated the mechanization of time, thereby preparing us for an artificial reckoning of time and forcing us to conform to an arbitrary mechanism. Thus, we became accustomed to rising and retiring at a fixed time, beginning and ending our working day at a specified time, performing our favorite activities at a set time. So we now have a designated time for sleeping, for eating, for working, for relaxing. Spontaneity has been replaced by regimentation. Indeed, time no longer serves us; we serve it. Mechanization has made us servants of time, but mechanization also means its very christianization, for Christianity liturgically shaped the significance of the moment by partitioning the day and consecrating its parts. Technology, as it developed in the West, adopts and extends this christianization of time, utilizing to the fullest the medieval concept of regulating and controlling time's mechanization.

Not only has technology made us extremely conscious of time, but its very presence has shrunk time to an insignificant lifeless commodity. Technology has made us so aware of time that we seem to have no time at all when we are most absorbed by activity. Time is now hectic, endlessly occupied, filled with excitement and confusion. Ironically, in the rare intervals when we are oblivious to time, we believe that we have an infinity of it because we feel its shapeless, empty amorphousness, like the time when we were children and felt that our childhood would be endless. Time-consciousness has made us crave a time we do not possess. Because we are so time-conscious, we relegate to free time all those activities we do not have time to perform, but even free or leisure time is so scarce as to be denied its rightful place in our lives. We do not so much use or manage time as spend it. A solution to this dilemma is the cultivation of idleness. Despite the fact that technology encourages us, against our better judgment, to perceive idleness as the absence of being, only work is perceived with its fullness.

Although we see our lives cut up into separate segments for the activities we engage in, life cannot be divided into fragments of time. Although governed by time, life is one continuous thread, as if it were unraveled, spun out from beginning to end. This continuity is what we mean when we speak of the temporality of being, defined by time's unbrokenness, because there is no being without time. And the same holds true for history, which does not deal just with man, but with man in time. We come to understand man and interpret him because there is time. Man's existence is fulfilled within the plenitude of time, undivided and indivisible. Thus, time does not mean just the present. It also means the past and the future. It means all the time made available to each of us, but time's indivisibility has meaning only within the limits of time, within its finity, beginning with birth and ending with death. It is no redundancy to say that to temporize time means to act in such a way as to bring being (and things) into fruition.

We are concerned with time, that continuum we experience but cannot observe, primarily because we have so little of it, because we know that all time is limited, both by our actions and by death. And knowledge of the latter drives us even harder to achieve all manner of exploits. We strive to conquer time and space, to run faster and faster, to fly higher and higher, to race about the earth against the clock, and all have meaning, as Ortega y Gasset wisely says, because we are mortal.[3] We experience time because we are aware of mortality. As for death, it is also influenced by technology's rationality, but negatively, because technology deprives death of meaning and robs it of any significance as an eschatological condition. Technology distorts death so as to give it a new qualification. Although it cannot disregard death as the termination of life, technology removes the death of the individual from one's own prerogative. The modern age dares to sustain life, when on the verge of death, by machines; it has the power to induce us to inhabit a body which is only technically alive. Life is prolonged not because we love it, but because we love technology more, and perceive it as a life-giving force. Not only have we cheated death of its meaning, but we have also degraded life by refusing to let it go.

Although death signifies that we are a part of nature, technology denies this relationship. Just as technology is loathe to leave anything in a natural state, it likewise attacks the idea that death makes man a creature of nature. Death, a key factor in the completion of existence, has been cheapened and falsified by technology so as to become yet another perfunctory phenomenon of a technological age. It is under increasing pressure in our time to serve no more than a utilitarian function. Most appreciably, we suppress mourning, which is too strong an emotion for a cold, insensitive and overly rationalistic age. If we mourn at all, we do so stealthily. And thus existence is robbed of the true meaning of death, and of the emotion that rightfully belongs to it.

More can be added as a criticism of technology. Not only does technology denude death of its existential meaning, it is also a cause of death itself. Technology's artificiality toys with and modifies the natural course of events, wherein (as one observes in the animal world) death occurs by means of predation, disease or starvation. To all these means, by which man too succumbs, technology adds another--death by decay of the living organism, which rarely occurs in nature. This is why cancer has increased its presence in the modern age, for cancer is a disease of progressive emaciation caused by the decay of the body through misuse, nondiscriminatory practices and bad habits, a disease of degeneration aggravated by poor choices in an unnatural world. Thus, technology both denudes death of its rightful meaning and causes death of a particularly terrible kind.

Notes

1. Underneath great changes lie men who are prone to greatness, that is, if we are to produce great changes, we must first produce great men. Hegel, *Lectures on the Philosophy of World History. Introduction: Reason in History*, trans. H. B. Nisbet (Cambridge, 1975), p. 84.

2. Kierkegaard, *Two Ages: The Age of Revolution and the Present Age. A Literary Review*, trans. Howard V. and Edna H. Hong (Princeton, 1978), pp. 100-103.

3. Ortega y Gasset, *The Revolt of the Masses*, English trans. (New York, 1932), p. 39.

Chapter 6

The Adulteration of Culture: The Impact of a Multitude

Each stage of man's societal development has been met with an increase in population. It occurred when man was a nomad in the hunting and gathering stage of evolution as well as during the first development of horticulture: each new human typology has encouraged a greater population density. Each technological manifestation--the use of fire, the domestication of plants and animals, and the development of tools--has done the same: there is essentially a reciprocal relationship between technology and population growth because rapid increase in population is both a result of technological progress and a cause of it. It is not only the effect of a higher standard of living and utilization of natural resources but the catalyst to social change. In the end, no matter what our views are of technology, the truth is that it has made us all too numerous. It has put far too much pressure upon nature and natural resources. Despite technology and the hopes that accompany it, the technologizing of nature is applicable only within very limited degrees, beyond which there is scarcity. Never before have we brought such pressure to bear upon the earth because never before have we pushed nature to such extremes. The tremendous increase of humans which has occurred in modern times is causing the earth's resources to dwindle. It causes environmental degradation and impairs the quality of life on earth. Ultimately, the spoils of progress are the limits of growth.

Thus, great masses of people enter the world through the aid of technology. The more there is technology, the more there are people, and for the most part, people are characterized by ordinary values, that is, ordinary people contribute little out of the ordinary. And because people are ordinary, their lives are not filled with great exploits or great ideas or great loves. Their lives offer little opportunity to profit from supplications of frustration, or innovations of occasional genius, since every increase in population signifies a greater acceptance of normalcy. The greatest agency of these great masses of people is the fodder to produce yet more people; but excessive population has no check upon its growth--an absence much aggravated by technology, which does not so much go out of its way to protect the ordinary as it creates conditions unfavorable to the extraordinary, unless, of course, the extraordinary become a part of technology's great scheme of things. Because technology serves as a stimulus to excessive population growth, a fact evident around the world, overpopulation intensifies the scourge of poverty for it increases alienation, crime and violence as well as the chance of death by starvation and disease.

Excessive population entails the collectivization of both people and institutions, for as population increases, the chance to govern falls to fewer and fewer people. Not only does overpopulation constrict freedom of choice, it also reduces alternatives and gives error a greater play in our lives. To put it simply, a smaller or stable population gives us more options. More to the point, overpopulation brings a lowering of intellectual acuity because the latter requires time, space and silence, all of which are lost to the teeming masses. We can all agree that mere numerousness of human beings has little or no value when considered apart from the quality of their lives. Hence, it is no surprise that the mass of men rarely reflect about their lives, but then how could many of them have time for reflection when want, deprivation and hunger gnaw at them?

The teeming masses give rise to anti-social behavior. Pushing and shoving, they are the cause of a new age of barbarism, but who are unlike those historical barbarians who militarily menaced the world. The new barbarians are different, for not only are they everywhere, but they are easily assimilated by the masses. These barbarians pull the masses even lower for they are their vanguard. Whether blue collar or white, they are in the forefront of mass morality. Poorly educated and lacking fundamental skills, they are callous in their universal disregard of culture, ignorant that culture constitutes the totality of all the manifestations of a people. Because of their shortcomings, they are enemies of civilization--uneducated, indifferent, unrefined. They accept the results of technology and its scientific basis without feeling the effects in themselves of the process of betterment which science and its technological application hypothetically induce, nor do they seem aware of the intricate historical process through which technology evolves. These barbarians are simultaneously self-abusive and abusive, self-destructive and destructive. They are harmful to themselves as well as to others. Unlike the barbarians of the past who were half-civilized before they were finally assimilated, today's barbarians are half-barbarized when we confront them. Rather than ascending on the ladder of civilization, the new barbarians are descending. Since they already are in our midst, the time is past when we can build a great wall to keep them out.

So there is a reason why any culture will falter if the quality of its population decreases in inverse proportion to its quantity. Therefore, a culture is vulnerable when it is reduced merely to a factor of quantification, when its specificity is subjugated to measurement. Fewer people who are better is desirable to a lot of people who are ordinary. Children born to parents who themselves are poorly educated, emotionally disturbed, mentally inferior or in bad health not only have a retarding effect on any age, but also lower the quality of people born to the next generation. Yet the pluralistic masses created by technology discourage ethnic and cultural identity, since they strive to overcome ethnic exclusivity by advocating mediocrity. In order to forestall this tragedy, distinct peoples and nations must be educated about their past, since a lack of self-identity obliterates the differences between cultures.

Of course, what we do is variable. Our actions change to meet changing social needs. But the natural environment also limits the material conditions of life, and it is nature that is limited in its ability to tolerate the changes brought to bear by us. Because technology gives us the power to exploit nature to an appalling extent, we have also acquired the ability to threaten our material well-being. Thus, the desecration of the earth means more than the endangerment of plants and animals. It also means that human cultures and human values are endangered. Since we are slow to acknowledge this condition as a possibility, we may be spared the luxury of making a choice when the reality of this possibility is presented as a demand. Of course, it is easier to destroy than to create, and certainly easier to waste than to conserve.

The multitudes produced by overpopulation, which are highly visible and disturbingly vocal, have acquired the ability to disrupt solitude. They challenge the individual's need to be alone, a principal manner by which we find ourselves. Understandably, the masses have become increasingly important in modern history, because man is frightfully lonely; that is, feeling estranged and cut off from the world, finding loneliness intolerable, we seek the companionship of others. But no matter how we look at the masses, whether we consider them effective as a threat to solitude or an aid to loneliness, they fuse all things together. Not only individuals, but all manner of things lose their exclusivity when appealing to the mass, for the masses are simply the overwhelming of bodies by sheer numbers. If this were not a world in which alienation is so commonplace, we would not so readily seek solace in the mass. Incapable of appreciating the need to look within ourselves, modern man shies away from himself. Indeed, the masses epitomize the modern predicament; they show that we are aliens in our own world, that we are isolated from others no matter how hard we try to be a part of them, and above all, isolated from ourselves. We have been relegated to the status of foreigners among an indigenous population.

Despite these shortcomings, there is a sense of joy for anyone who feels that he is part of the mass, but it is a negative sensation because it is exultation over the denial of freedom, delight at the flight from responsibility, frolicking with indifference in the face of discrimination. Indeed, what passes today as the well-adjusted individual is not the man who struggles to be different without being eccentric, but the mass man who conforms, who is smug, complacent, happy with himself, doing nothing, going nowhere--the very epitome of an anti-man. Since we are actively encouraged to identify with the mass, we also associate our desires with it. But more than this, however, we feel that the mass really belongs to us, that it functions for us, that when it acts it is our mass, expressing our desires, completing our actions, even thinking our thoughts--if we had thought them. Hence, we can understand the overwhelming influence of the masses over the individual, what Heidegger calls "the dictatorship of the they."[1]

Who would deny that the burden of technology which we have hoisted upon our backs bears down heavily upon us? Although we all must shoulder some burden in our lives, just as we shoulder the weight of the past, we must not neglect to bear the weight of our humanity. But to the man in the mass, there is a strange indifference to the suffering of other men, a great unawareness of the rest of humanity. Although physically close, they are humanly remote, signifying distance from time as well as isolation from history. They are removed, abstracted, inaccessible and unapproachable, stretching out into a vast wasteland of insensitivity, dehydrated by the blasting sun of disinterest, entangled by the vines of indifference. Since the masses are a degradation of the human race, they express their humanity in a debased form. And because the masses associate too closely with the values of technology, they run the risk of feeling too much at home in it. They find their identity only in and through its benefits. We all feel this way at one time or another, but how dreadful it would be if we felt compelled never to trust a dissenting opinion, never to remove ourselves from our familiar orbit. Technology envelops us all too completely. Wherever we go, wherever we step, it is there. And mass media, technology's voice, constantly reassures us that we are able to solve all types of problems, from bad breath to anxiety, from constipation to loneliness, and that we may do all these things easily, quickly, and without effort, in fact, almost without our participation.

Nor is it surprising that there is so little real political struggle in an age that surrenders itself overwhelmingly to technology because politics on the grand scale, when individuals organize and oppose the established order, are rendered meaningless, since technology proposes to do everything for us. Above all, it becomes the spearhead of the democratization of the world; that is, technology becomes the agent of the world's mediocrity. But in addition to technology's presence, the masses themselves lend a hand to their own impotence--because no matter how large the masses become they always constitute a vacuum. Their pure numbers denude them of their strength. Again we must remind ourselves that to reduce a people to a mere factor of quantification is to emasculate them. For surely, the masses have demonstrably less control in a technological age than in former ages, causing their feeling of powerlessness to give rise to the feeling of victimization. But more than this, we can no longer exert great influences in the world because we are no longer intense; we are no longer soaked deep with the world and our unique place in it because our passions have been extirpated, pulled out at the roots. The driving forces of times past, when man was saturated by the feeling of mystery and wonder, when he was surrounded by the unknown and unknowable, have left him. Now, everything is explained, everything analyzed and dissected, everything laid bare by probing technicians. Nothing anymore is left in its natural state.

When it comes to politics, the masses have been reduced to expressing their views through political parties, because the latter, which first appeared in the nineteenth century, are the only means the masses have for political ends. With the party system, the masses have lost more than the importance of individuals in the political arena; they have also lost qualities of merit, integrity and excellence in the individuals they elect. The people the masses elect to office have no saving graces, except the distinction of being so colorless as not to antagonize anyone. Running for office means, more and more, manufacturing a public image attractive to everyone, regardless of what a candidate stands for, making the ideal candidate that man who gives a good impression of himself to everyone, who hides his true feelings when assuming the role of another, thus, an actor--and when actors become elected to office, even to the presidency, they become very well suited for drama, or, to be more accurate, melodrama. Thus, politics in a technological age brings showmanship.

Since the spirit of the masses is perceived to be the spirit of our time, whatever is the newest in contemporary culture is that which sets the norm, that which we are expected to follow faithfully. But we should resist such an outrageous statement; we do not have to be passive in the face of cultural and historical forces, since we exist in a dialectical relationship with them, for all of us are links in the great chain of being. We must remind ourselves that we are capable of shaping an age's institutions as much as we are shaped by them. We can put pressure to bear upon them just as they press us ever more tightly. We can project unto the future just as the future bears down upon its past who we are, where we have been and that to which we are going as if we are capable of moving both forwards and backwards simultaneously. Rather than to be someone, especially the someone that mass culture advocates, we should become someone, to become the masters of our destiny and the decisive force in existence because both the world and ourselves influence one another, because both are juxtaposed to each other. We need to make our influence greater, to stand out from the crowd, to be distinguishable from all others.

But until we come to realize our latent possibilities despite the pressure of the masses, we fall victim to a process of atomization, which is characterized by the lessening in importance of the individual so that everyone becomes merely a minuscule part of the mass. And the masses are omnipresent. Whether at home, at work, or at play, the masses are evident everywhere. Wherever they go, their tastes are alike. The masses eat the same type of food, seek the same type of entertainment, wear the same type of clothes, are influenced by the same type of advertising, manifest the same type of fears. They are constantly told to desire what everyone else wants, to be one like many. Indeed, the masses want things because other people already have them. And they are easily swayed from one option to another.[2]

The masses cannot be sustained without ideologies. They must be constantly told what they must do; or if they wish to be slaves, what they must not do. And ideologies that characterize the masses, like all ideologies, must be supported by an elaborate artifice of propaganda, by those ideas, facts and allegations deliberately and publicly circulated in order to attract them to concepts and causes considered to be favorable or to dissuade them from those considered harmful. Ideologies are always sustained by assumptions, which when no longer believable are easily identifiable as lies. If we are distrustful of the description "propaganda," then let us call it advertising; but all advertising is laden with hidden meanings and innuendoes, promoting some goal or ideal when compared to others. And, of course, it exploits the gullibility of consumers, which is intensified by mass media. In fact, the press, which is one form of media, has become the conscience of the masses because it makes itself the tool of public opinion. Although the masses are manipulated by these forms of propaganda, they are largely responsible for today's cultural emptiness for they take the final step of accepting or rejecting its message.

Since advertisements appeal to shared experiences and feelings, they are communal phenomena. But rather than imposing a mentality, advertising reinforces it, that is to say, advertising or mass media reflects more where we are than where we are going. It is a response to morality, not an attempt to shape it. And like many other institutions, advertising fosters illusions, the principal illusion being the need to conform. But advertising also attempts to reduce its audience's attention span to smaller and smaller units, which are fragmented and isolated from each other. If it achieves success, its audience will find thought itself a most arduous task, rendering the masses even more inarticulate. As life becomes more subjected to the masses, it will become easier to advertise. It will become easier first to invent something and then to create a need for it, for we have needs which are socially produced, that is, needs which are tied to and satisfied by other people or by social institutions which render these needs possible. Because of the interplay between us and the social pressure created by others, some of these social needs may be transformed into demands, into desires normally satisfiable by means of coercion. Thus, people may demand certain things simply because a need exists.

In conjunction with the masses and the needs and desires associated with them, it is evident that the number of people who actively participate in a culture is a matter of great importance because as a culture becomes more ordinary and less threatening, it can support more people. It does not matter how large a host becomes as long as it remains compliant. Although rural cultures, such as China and India, are capable of supporting large populations, cultures tied to technology as a means, which are always urban, also have this capability. Apart from the presence and encouragement of Western style liberal politics, rapid population growth is the result of the perfection of technology and its influence for promoting life, above all, in technology's democratization, so that now we number in the billions.

Truly, the modern age is the age of the masses, a time when the strength of pure numbers has acquired new meanings, a time when the great weight of so much humanity bears down upon history. And whether or not the masses will leave a legacy remains for the future to determine. But let it be known that the masses, regardless of their origin, are forever bound to technology for the continuation of their survival. Technology and technological progress are their lifeline. This is why the masses are irretrievably tied to technology because just as technology creates the conditions which give them birth, any decline in progress means their own decline. Such a dependence is a most precarious one, offering little protection against the unforeseeable. And with the masses comes its own type of culture--a culture of pure numbers, but denuded of depth or dimension.

The type of culture created by the masses, which is essentially quantitative in substance, always aims to exploit the desires of the mass, which more often than not are expressed as diversions or dissipations, and is usually conferred by those who are in some sort of control over media. Much like traditional morality in general, mass culture is imposed upon those who are influenced by it. Because it comes from above, or to be more accurate, from outside, mass culture is not to be confused with folk culture (the culture of the people), although we are living in a time when the latter is rapidly disappearing. Above all, mass culture thrives in a pluralistic age because the latter assimilates all oppositions, all polarities, all antagonisms into a common civilization, which manifests unity only externally. Since a pluralism, because of its democratic basis, is believed to express the general will, it is considered to have an appeal to everyone, and what appeals to everyone are banalities, those factors devoid of distinction and refinement. Mass culture aims at bringing everything down to a level so base that no one would be envious of another because to reduce anything to the lowest common denominator is to obliterate it. And all mass cultures bombard their public with cultural minutiae, with microscopic bits of information so minuscule to fleeting faddism as to be insignificant, and so insignificant as to be worthless. Thus, the notion that culture is the totality of all the manifestations of a people has little meaning when these manifestations are so unimportant. They are merely fillers of time, made up of petty diversions that are quickly forgotten because they leave no lasting impressions.

However, culture for the masses is not culture in the usual meaning of the word. It constitutes a pseudo-culture, an overly simplified and therefore bastardized imitation of culture fashioned for consumption by the consumer, since it is preoccupied solely with the marketability of those ideas within its sphere of influence. Above all, it does not require discrimination in judgment between what is better and what is worse, but an ambivalent indifference to both, because mass culture can be either sacred or profane, prudish or pornographic. Culture for the masses cannot produce any art form which may be called classical, that is, the art form which is both uncompromising and demanding. Although we must expect a great deal from classical art, it also expects a great deal from us. Followers of mass culture are more likely to know the names of Hollywood actors than the names of Virgil or Dante or Goethe; and this is because Hollywood belongs to mass culture, but the latter do not. Because mass culture is incapable of setting standards of excellence, it limits itself to what it can do, to setting standards of popularity.

For this reason, the simple is given greater credence than the profound, the frivolous greater significance than the serious. So rather than classical exploits we have trivial pursuits. What is quick and simple is more to mass culture's liking than what is long and involved. In fact, mass culture, which ironically parades itself as artistic, cultured, tasteful, discriminating, creative, noble, is really hostile to art because it loathes life, because it is revengeful of will, defiance and choice. Mass culture is conformist culture. By comparison, genuine art and culture by its very nature cannot speak to or for the masses, because by definition, genuine art and culture are exclusive. Genuine art and culture are aristocratic; they are created by a minority of people, and appreciated by a minority who know their gift. An understanding of genuine culture demands effort by its patrons. The so-called culture generated by the masses is little more than a means of distraction, as when we are more likely to look at a woman completely naked than if she were elegantly clothed. Mass culture draws our attention away from what is meaningful. It replaces the serious with the sensational, and the sensational is what sells best because what appeals to a mass audience is vulgarity. Thus, mass culture absorbs the masses; it aims to make all similar, none different. Since everything and anything is compatible with mass culture, so typical of the age we live in, it includes everything, that is, it makes everything a part of itself. Like cancer, mass culture incorporates its surroundings, spreading by invasion and other devious means. And because mass culture consumes everything, because it is omnivorous, it cannot help but to diminish whatever it encounters. Mass culture dilutes whatever engages it, attenuating the world at large.

And for a similar reason, mass culture is most notably visual. It emphasizes pictures rather than words, that is, orality rather than literacy, sound rather than symbol. Being preoccupied with images, black and white will no longer do for us. We must see everything in living color. Not that we must comprehend all, but we must see all. We must reduce everything to the sense of sight, emphasizing the values of the body over the mind. Because our time is so visual, appearances are critically important. Thus, we are so racially conscious, so fashion-minded, so very well aware of physique, so given over to youth. And because technology abhors secrecy, we are concerned with the bare essentials of our emotional and physical constitution, exposing the most private aspects of ourselves. Nothing remains hidden, for everything is scrutinized by the naked eye, everything given over to voyeurism. But seeing everything is limiting. There must be some part or some aspect hidden, draped, clothed. There must be illusion or the resemblance of it, because illusion stimulates us to explore beneath its cover; it incites us to delve, to probe, to enter.

Although artificially fabricated, mass culture is not created, but produced, not made from life, but manufactured for the sake of novelty, for the blind adoration of fashion. Although mass culture comes from a culture, it does not completely belong to it. It is a synthetic product. It does not flow naturally from life's genuine cares and concerns. And wherever culture is denuded of sincerity, it is denuded of substance. Like Hollywood, mass culture is the means by which we escape from reality's harshness. And that is why, once manifested, mass culture is always before us. There are those among us who find it hard to believe that there was a time in the not-too-distant past where people lived without television or automobiles or computer games. The absence of mass culture gives the incentive at least to confront reality unpretentiously, like one standing naked before the Almighty, standing alone without support or diversions, paying attention to what is truly meaningful. Mass culture, on the other hand, is merely a filler of emptiness, an expletive, like shopping, because one does not know what else to do.

Similarly, what passes for literature in a mass culture is quite removed from its conventional meaning because traditionally literature has been defined as the attainment of a perfected form in prose and poetry that concerns those things of lasting value to the human condition. In a mass culture, literature does not strive for excellence, but for mediocrity. This is why it appeals to the masses, why there is such a description as best-sellers for the mass market, and what sells best is what sells often, so frequency of sales, not content, defines best-sellers. Consequently, we now have romances, which take us nowhere but into the expectant arms of the beloved, or science fantasy, which takes us right out of the only world we know. This literature is not elitist, but egalitarian, not aristocratic (intended for the best), but democratic (appeals to the people). It is based on the assumption that anyone can read it because it says nothing. Hence, it is trimmed to the tastes of an increasingly lethargic public. And because it says nothing of lasting value, it is of inferior quality, that is, it concerns itself with trash. In fact, it survives on it. Although our age takes great pride in its ability to communicate, it is characterized by illiteracy. Our age is suspect because to be truly literate means having a knowledge of the past, of one's culture. Being able barely to read the newspaper does not make one literate. In fact, literature itself has become degraded to mere entertainment and little else, since the masses would rather laugh than think. Literature and all art forms are rejected when they exceed the standards set for them, when they give their audience more than it is accustomed to absorb. Hence, the disappearance of genuine art and culture occurs when the world finds them repugnant to their tastes, superficial to their needs.

A technological age throws into question all forms of genuine culture. Since lasting culture always appeals to individuals because it is for the individual that it is meant, it generates a one-to-one relationship with those it enlightens. If a multitude is moved by it or understands it, so much the better, but its appeal to a mass audience is not innate in its intent. Its appeal to a multitude is a consequence of its presence, not a cause of its being. In comparison we can say that individuals have no direct relationship with each other when they are members of the masses. They have a relationship only to what makes them a mass. Thus, the masses are created by severing all interpersonal ties, and replacing them with a common goal to which everyone swears allegiance. The growth of the masses infers the subjugation of the individual, that is, as the masses grow in power, the individual diminishes in importance. The masses, which are homogeneous and compliant, objectify its members. And this is how the masses and technology have evolved into a reciprocal relationship: the masses foster the acceptance of technology by laying a basis for it, and technology creates masses by subjugating individuals. Truly, we live in a time when individuals no longer have meaning, when groups or assemblages of individuals acquire meaning only as abstractions.

Nevertheless, culture has always been and will always be influenced by technology, nor should we ignore this fact because culture is a technological phenomenon. Technology is also made possible because there are masses. If there were no masses to take advantage of technology, there would be little need for it. A mass culture is always a technological culture; the two entities rise and fall together. At one time the masses, in so far that they existed, were subjected to a creative minority, but when this occurred, the masses were quite small. Now when the teeming masses have become so vast, this creative minority is rendered impotent to create culture or is encouraged to produce a simplified culture for the masses. Although both individuals and masses are needed, neither should be completely overwhelmed by the influence of the other. What is culturally different in a technological age, so unlike all former ages, is the influence of the masses on culture, necessitated by the magnitude of their presence, and the profound impact of technique on modern cultural forms.

Since technology obliterates the underpinnings of the lower levels of culture and depresses its higher levels, culture consequently sinks downward. What remains is a homogeneous middle level of mass culture, understandable by almost anyone in touch with the rudiments of society, but signifying nothing really meaningful. It is unfortunate that when art and culture lose their aristocratic basis, that is, their grounding in the highest standards of excellence, they lose their basis in fact. This is to say that technology causes cultural disintegration; it strips culture of its meaning and leaves it vacuous. Although it is difficult for an individual to create art and to be recognized for it in any age, it is even more difficult when that age is technological. Individuals who are visionaries of their time but are ignored because they disagree with conformist art are surpassed by other better known contemporaries. Such visionaries are considered by their contemporaries to possess no useful skills for the betterment of humanity because their skills are regarded to be of little practical value. The advancement of technology poses an ever-increasing threat to the perfection of genuine art, for not only is the latter difficult for any age, it is becoming increasingly arduous in a technological one. Of course, more commercially-conscious artists, bridging the art galleries and commercial advertising, blending high and low art, will always find it easier to achieve recognition than those whom later ages hopefully will call its real and lasting representatives.

There is mass culture because there are masses. Although there were semblances of the masses in the past, they exerted little or no influence on the culture of their time. Only with the rise of industrialism in the eighteenth century has world population soared and with it the masses' influence, but the masses are not really composed of mass men, but of individuals in a mass. The mass man as such is merely a figure of speech, who will remain mythical as long as he has a private life, that sacred area which keeps the complex, technological world at bay because a private life is an enclave of free choice. But a private life is no hidden sanctuary because it is invaded constantly by advertising, peer pressure and the discomforts of nonconformity. Unless one is strong, little remains. Even if everyone had exactly the same aspirations, the same hopes, the same desires, there still would be no mass men, because each self is exactly that--a being unto itself. The principal argument against the masses, simply put, is that they destroy cultural values, that it not only makes no distinction between the exceptional and the mediocre, but that it makes the exceptional vulgar and therefore degrades it. It seeks out the common at the expense of the unique.

And the groundwork for mass culture is mass education. Although education is intended to enlighten and broaden, often it is informative only on a very superficial level. The critical apparatus needed to develop all one's faculties, which comes with a classical education, is denied to the masses because the latter have neither the patience nor the inclination to study a rigorous curriculum that offers little practical value. The remedy to this estrangement is technical education, not education which stresses the humanity of each person, but education which emphasizes the involvement of each person in a feigned and fictitious artifice, education of a purely meddling and distracting nature. The emphasis on practicality makes definite economic sense, but no other sense. The existential knowledge of who we are, the development of self-discipline and critical independence, the awareness of our own imperfections and above all, the consciousness of freedom and the burden of its responsibility, are all lost to a technical education because the latter disrupts the sense of balance between the individual and the social and physical worlds, creating a state of abnormality in which individuals no longer have continuity with the world (and its past), but are juxtaposed with it. This disequilibrium or lack of tradition means the absence of any feeling of devotion to genuine and lasting culture. And this lack is what characterizes mass education, which reinforces the discontinuity of mass culture, education which is anti-intellectual, materialistic and only superficially beneficial.

But there are other thoughts, too, about technical education that we should consider. Since technical education promotes itself as having the best approach to the acquisition of knowledge and because it promotes "useful knowledge," it regards all other knowledge as useless, or secondary at best. So technology is concerned with facts, not theory, with information, not speculation. And facts are applicable, concrete and precise, readily interpretable, easily manipulable. Technical education argues that if speculative knowledge serves no technological purpose, it should be replaced or left behind. In as much as technology strives to manipulate the present so that it may alter the shape of things to come, technical education exhibits itself as modern, useful and fashionable. It stands ready to attack and transform nontechnical education and all forms of education that deal with the past--not only the sciences, but the humanities as well. All research today is technical, and concerns itself with technical points of view. Although technology aids the pursuit of learning, even when it deals, for instance, with prehistory by perfecting the technical tools and instruments of research, its adaptation to those times long dead and forgotten is purely coincidental, because technology's influence cannot help but spread everywhere, leaving no stone unturned.

These views also tell us a great deal about the place of art in our day. Since art should always manifest life, this representation is jeopardized when art is consumed totally by technique. Because the latter proceeds at its own expense, the increased use of technique removes the artist more and more from life when he creates art because it distances him from being. Although both life and technique must be present in art, technique must always be subservient to life, otherwise art would not be real, but artificial, not succinct, but diffused, not natural, but contrived. A truly revolutionary school of painting in a technological age would be realism. Abstract art forms emphasize the emptiness of the technological age in which they are produced, emphasizing style or technique over life. They demonstrate not only that modern art is a form of protest against tradition, that is, against the art of the past, but that modern art itself is produced for other artists, for other rebellious souls, who absolutize their own rootlessness. Abstract art has become equivocal, indecisive and questionable, which, of course, it wants to be. This growing aridity of art is caused by the deification of technique. Art no longer is representative of life, but of itself, if you will, art is now for the sake of art. The debasement of art, the preoccupation of technique over life has dehumanized art. Because art now misrepresents life, it destroys the hope of communicating with it, and that is its genre.

Thus, we live in a time where there are three manifestations of art. The first of these is genuine art which is created in conformity with its time, which appeals to everyone and no one. This form of art is classical both by what it portrays and what it demands. Then there is art for art's sake, art produced for other artists, art as exclusive as genuine art, but produced in rebellion to the age in which it is made. And it is art indifferent to its viewing public. Lastly, there is art for the masses which lies between these two. Art for the masses is the meeting ground for genuine and technical forms of art. It is the gray area where genuine art is adulterated and technical art is marketed.

The genuine artist who creates art for life and not for art's sake need not create art for everyone, but everyone he does create for must understand the rarity that is creativity. Increasingly, this artist is a lonely man because those for whom his art would have lasting meaning have diminished to a paltry few. Not only is he a stranger to the multitude, but the cultural heritage that he has inherited has been diluted by a prejudice which says that we no longer need meaningful art because the purpose of the latter conflicts with technology's purpose. Although conditioned to technology as the means of its handiwork, genuine art remains essentially a nontechnical endeavor. Of course, this evaluation pertains to the genuine artist, who is either invisible in his own time, or visible, but ignored, and who is to be distinguished from his counterpart who prostitutes himself for commercial success.

Art, like love, cannot be contrived; it cannot be predetermined because it is not wholly conscious. And although art is shaped by the manners and mores of the times, which includes technology, a heightened technology is a hindrance to true art. Not only the passage of time but the very advancement of technology severs links with the past. Thus, like time, technology is a destroyer of culture. When any age develops one type of manifestation, it loses another. As an age gains one insight, another slips away. Thus, we cannot do things as they were done in the past because we have lost the milieu which produced them, since each age has its own vision. This is why one age looks different from another. But these thoughts do not imply that art, like morals, evolves. What we are saying is that although technology progresses, art and culture do not. Art through the aid of technology makes advances because of accumulated knowledge and technique, so that now, for example, the art of painting in our culture uses not only oils and water colors but computerized graphics, all of which were acquired at different phases of its development. Because art and culture are directly conditioned by an age's intensity, they are dependent upon its spirituality, tied to its vitality, fashioned by its self-awareness. And none of the latter phenomena can be recreated in any age which has passed away, since art is created in relationship to the totality of one's culture because all art is metaphorical, symbolizing the values of an age. There is no such thing as art created in isolation, for even art created in a cave is never devised by men living in seclusion.

Art is expressive of an age, reaching out more and more the deeper it digs within. This is why impressionism and cubism are legitimate forms of art because they accurately depict the lack of clarity noted for their time. But abstract art is not a legitimate form because it completely circumvents the responsibility of art's calling, because it burns the bridge needed to span the chasm between form and content. Nevertheless, art can be created in rebellion to one's age, in defiance, in protest to the very history of art, by seeing all previous forms of art except its own as exclusive, created for and appreciated by a special group of people. And when we encounter this interpretation, this technical manifestation of art does not even speak to the masses, for they too find it incomprehensible. This is what we mean when we say that what passes itself off as art today is created for the sake of art only, that is, for the sake of artists only, who are really little more than technicians.

Apart from its technological basis and the movement of time itself, great art in the past has never been in conflict with the age from which it sprung. A century ago it would have been inconceivable for an artist to oppose his culture, to renounce his own traditions. Today, artists struggle with what they find in the world, antagonistic to and alienated from the values of their time. Although indispensable, the individual must not become a bane to art. He must not relish his own isolation, his own creativity, his own subjectivity to the exclusion of everything else. The highest forms of art are created because the artist sees himself as participating in something greater than himself, taking part in a quest transcending himself. And the greatest hindrance to true art is the rigidity imposed upon it by sophisticated technology and by the latter's excessive rationality. True art must be both free and limited, spontaneous and conditional, subject to technique as the means by which art is created, yet uninhibited by it because true art is no science. To be meaningful, art must assess its end in the light of its means.

Simply by definition, art and culture in the sense of refinement are always elitist. They always attract and are attracted by the few. As for mass art, what appeals to the masses is either popular culture or perhaps high culture deliberately vulgarized and debased. Or to put it another way, genuine art in its true form is not accessible to everyone because it cannot be made understandable to the vulgar. When art appeals to the masses, what they find appealing in it is a mere characterization of art, an illustration or portrayal of imitations. Because the masses constitute the greatest number, in fact, the whole of humanity, they produce simplifications and stereotypes specifically designed for everyone in general, but no one in particular. The masses represent a common spirit, or to be more precise, the lack of one. For here is proof that genuine art is always produced by creative individuals or by creative minorities. Because true art is no science, Mozart's symphonies are inconceivable without Mozart as Donatello is indispensable to his bronzes, although the discoveries of science can be made by anyone working in a laboratory. True and lasting culture is still an area where only the few are meant to tread. Unfortunately, culture reserved for the few and never intended for a multitude is now invaded by the masses who have gotten a taste of high culture. As we said above, mass culture spreads like cancer, consuming everything in its path. It subjects everything to the same weakness; it belittles whatever it encounters.

Notes

1. Heidegger, *Being and Time*, trans. John Macquarrie and Edward Robinson (New York, 1962), p. 164.

2. Jaspers, *Man in the Modern Age*, trans. Eden and Cedar Paul (New York, 1933), pp. 39-40.

Chapter 7

Lost Among the Stars: The Secularization of Religion

One would think that all of us seek the meaning of our lives, given the time and temperament to do so. Although ultimate questions are asked by all peoples and all ages, people in the past seem to have had a solace that we lack: they believed that there was a God who cared for them, that they could find answers to the riddle of death, that they could achieve a measure of significance in the vastness of the universe in which a divine mystery inhered. What makes the modern age so different in this search for meaning is the feeling of despair aggravated by our abandonment of the idea of God's existence--what Nietzsche called the death of God--although the Copernican revolution endangered the relationship between man and God by removing man from the center of the universe, a dislocation which in time was intensified by both Descartes' concept of consciousness and Kant's concept of knowledge. Certainly, the modern age is an unfortunate time to be desperate because the feeling of helplessness runs deep, the sense of loss overpowering, the quest for meaning ambivalent. And because answers to our questions are hard to find, many seek comfort in traditional religious forms, or in fundamentalism which is alien to modernity.

Nevertheless, there are those who believe in nothing and bring nothing into the world, who idle away their lives with frivolous activities and diversions, devoting themselves to their jobs or careers because they have no other devotion. When removed from the customary routines of life and when given the opportunity to do so, we seek a more fulfilling definition of ourselves through work, because work is tied to the quantitative conditions from which it is derived. But we are all very much preoccupied with activity that none of us really has time for anything. We are kept busy so that we cannot reflect about ourselves or the world around us. Not only do we deify work because we have too much of it, but the very pace of our age leaves us little time for introspection. Thus, to some extent the present age has replaced spirituality with occupation, thereby making work a substitution for prayer. Perhaps the present age has already made its choice between belief in something beyond itself and belief in something in spite of itself.

But this is to be expected. It is expected that we would pull away from religious feelings in such a spiritually baseless age, for that is the nature of the age we have created. It is expected that the two principal problems encountered by religion--explaining the inexplicable, and overcoming or at least justifying the fear of death--can only now be confronted in a most inadequate and unsettling way. When true and complete answers are lacking to the questions that afflict us, we refuse to seek them out, and instead opt for fictitious ones. Ironically, this uncertainty is itself the situation into which religion makes its move. Mystically, religion arouses us, reaching out beyond human experience to tap the unlimited resources of a divine presence, which is believed to transcend all knowledge, but intellectually, it puts us to sleep, or at the very least, indoctrinates us to empty platitudes of meaningless rhetoric offered to give consolation to those who have a need for such limitations.

Although technology cannot answer the questions of life and death, being and nothingness, God and the cosmos, this inadequacy does not prevent technology from being molded into a religious form. Technology is not prevented from devising a way of life, nor is it prevented from demanding obedience to its commands. It even advocates the coming of a certain type of kingdom, stressing the infinite value of progress and the righteousness of the commandment of scientific inquiry. It demands renunciation of the old ways as wrong and acceptance of the new ways as right. It categorizes nonscientific writings as apocrypha. It demands recognition not only as a rite, but as a will to a theology, in a word, as a faith. And by accepting technology, we renounce belief in false idols, giving up our enslavement to iniquity, repudiating customs offensive to technology. Thus, technology has become the religion of our age, and has all of its characteristics, including proselytism. Its rites of passage are evident in its techniques, its priesthood in technicians and engineers, its temples in the halls and academies of science, its catechisms in technical journals and manuals, its dogma in scientific rationality, and its sacred history in the idea of progress.

Because technology is oriented toward the future, it has little regret for the past. To all believers who profess faith in technology as the only truth, its priesthood, that is, its technicians, have become ecclesiastical authorities, authoritative interpreters of sacred texts, who encourage believers of other faiths to apostatize. These priests stand ready, like all priests, to administer to that one important aspect characteristic of all religions: belief in a manifesto which puts up barriers to the self, that is, belief in something beyond oneself, to a conviction thought to be greater than the self. Religion displaces reality, much like fantasy. And when we believe in religion we do not believe in ourselves. We accept it without criticism, without in the least standing back and looking at it fairly. To struggle against religion is to struggle against someone else's perception of the world, to contest a mode of insight like Hegel's obsessive system where everything is meant to fit with utmost precision. Since religion is whatever we give our highest respect to, there is no other entity respected more in our time than technology. And like traditional religions, technology does not voluntarily alter its doctrines, for even when threatened, it never goes on the defensive, but seeks out offensive positions. Like religion, technology is seen as the prerequisite for the salvation of man, the means by which man is "saved" from himself.

Although religion is not confined to a single, individualized idea, but to a sum greater than the parts which go to make it up, it is always tied to that organization which is followed by varied and intricate beliefs. In comparison it is apparent that technology is a poor example of religion because it is monolithic, because it lacks the plurality of beliefs, and ironically because it is secular. It also contains no magic and no mystery, and lacks the notion that religions are considered to be based upon sacred writings, divinely inspired. Nor are traditional religions, unlike technology, reducible to one idea, because an established religion consists of a plurality of ideas, each interlocked with all the others. It is not so much that science and technology are opposed to traditional religions; rather, they stand against religious speculation and wonder, against revelation without dependence upon immutable laws and irrefragable truths, against a divine inexplicable being who is unattainable and defies comprehension, sitting upon a throne, flanked by his supporters. Traditional religions are not displaced by science or technology because the former are institutions well-endowed with bureaucracy that itself possesses its own techniques, but because they are no longer considered to be true, no longer regarded to speak with a voice worth hearing, no longer believed to be the only path to divine grace. The more science and technology become esoteric, the more they remove themselves from the masses, particularly what the masses regard to be truly meaningful, even if such beliefs appear base and banal to the sensitive. Therefore, traditional religions, which are now less literal, have little to fear from such rivals, even when science, which is now less materialistic, seems to deprive religions of their usual explanations of the nature of the universe and the origin of man. What they have to fear instead is a degenerating indifference, sustained by boredom, despair and disenchantment, whose only redemption is salvation at the hands of a pharisaic ministry.

It cannot be denied that we have passed through an age of transition which had awaited the birth of new gods. A new god has appeared, and it is technology. One must admit that there is some irony in the fact that technology associates itself with religion, that although traditional religions have failed us, the notion of religion has now become the soul of technology.

That transformation has come about because technology elevates its own view of reality, of reason, of organization, of stability--while simultaneously suppressing impulse and instinct--into a matter of religious experience, and the more technology evolves the more evident this tendency becomes. Like religions in general, technology is both spiritualistic and moralistic. Although there is a relationship between the two, religion is not the sole basis of morality.[1] Locke's arguments were never true, but at least all moralities do have a basis in fact, that is, they have validity because they concern how men live and interact with other men. But despite these observations and despite technology's attempt to alter morality, technology remains an inferior type of religion. Although religion gives a feeling of dependence in the sense defined by Schleiermacher, it is a dependence founded upon belief, and although technology also fosters dependence, its dependence is tied not to a hidden realm of being, but to its overtness, a dependence that renders technology into a dispassionate, insistent and systematic god, a god we are afraid to depart from, a god we feel obligated to cower before. So both traditional religions and technology tie man to mechanisms external to the self.

Although transformed into a religion, technology has called into question traditional religious forms, those categories of thought that we are most familiar with and that appeal to the most basic feelings within us even when they no longer suffice. The technological age has certainly questioned the relevance of Christianity. It has called into doubt that great cultural force which antiquity gave to the world and which the Middle Ages nurtured into its full flowering. Technology has overridden Christianity just as the latter supplanted paganism, but there is an obvious difference between these religions. Although the rigors of Christianity have fallen away, no matter how much reverence it may be given because people do not know what else to worship, it has had a most important and profound impact on the present age because our age would not have perfected its great emphasis on technology without the help of Christianity. To put it simply, technology is the child of Christianity and a scientific, Christian Europe. And technology's past experience goes a long way to explain why it takes such a linear view of the world, why it lends itself very easily to exploiting and dominating nature, why it proselytizes with sword and book. Although Christianity has fallen from grace, hopefully its metaphysics will survive, the metaphysics which depicts an individual, apprehensive self-doubt before the absolute, a solitary self before the infinity of being. Hopefully, we will go beyond the obsession with a once transcendent God who has now allegedly revealed himself to man.

Nevertheless, the waning of traditional religions ushers in technology's fatalism, that it can absolutely determine the course of the future, that everything can be planned, controlled and regulated to the smallest degree. If we idealize technology, it is an ideal we can know and understand, that we can direct to do our bidding, that we can manipulate. But such a notion is opposite to that which says that man is predisposed toward belief in the unseen and unknowable, corresponding to the Greek idea of fate. This is why in the last resort technology is a poor example of religion: because it strives to erase life's mystery, because it gives us nothing after we accept it, except more of the same. Although traditional religions have created a transcendent realm to the abandonment of everything else, technology in its opposite mode is no better. Traditional religions have advocated that because our bodies confine us, we must let our spirits soar. Yet, religions as well as technology refuse to accept man uncontaminated or untainted as a legitimate realm of influence. Both manipulate and exploit him because both distrust him. Both see him as morally impure, repulsive and contemptible, and both try to make him clean, simple and obedient.

While technology is rational and voluntary, traditional forms of religions are irrational and involuntary. Accordingly, the excessive rationality of a technological age must be contrasted with the arbitrariness of religion. Naturally, there are fundamental differences between traditional religions, which look to the past, and technology, which theoretically, at least, looks to the future. Traditional religions mimic what has passed away because they concern the attainment of a life that has already taken place. This worship of the past is called tradition, and tradition is merely a longing, not a way of life. It is little more than sentimentality, an admiration of what has been. Technology, on the other hand, is the opposite because it looks to the future. Since religion is only one type of experience common to mankind, nevertheless it is closer to those social truths, the truths that relate to others, than to those ontological truths, the truths of the self. But replacing religious values with technological ones has not eliminated the need for convictions because just as the man of faith is tied to his convictions, and therefore finds it difficult to make concessions, so too is the technician.

The description of religion as a means to power fits well with technology's definition as a religion. But power to be powerful needs continual reassertion; without reassurance it would simply dissipate like fog rising above the earth. Power, and most notably technology's form of power, is a means to an end, never an end in itself. Not only has technology evolved into a religion of power, it also manifests the power of religion. Since power is conditional, it is limited to and defined by circumstance. Changing situations automatically change the conditions of power. This is another reason why technology seeks stability because the latter confirms the maintenance of control. However, we now encounter the contradictory fact that technology in no way wishes to introduce real, substantive changes into the world, changes that might so modify the world as to be a threat to its very existence. Ultimately, technology wishes merely to alter the shape of things to come, not change them, since to alter means making things different without changing them into something else.

Because technology is both omnipotent and omnipresent, it dictates to us what is considered piety. But the shortcomings of technology as a religious form are most evident in its application by the masses. Since the masses are rendered possible because of technology, technology creates the situations which aid their growth. Despite the seemingly favorable conditions of better health and increased production, which we have already discussed elsewhere, technology also engenders apathy and indifference. But the characteristic which most accurately epitomizes the masses and is most characteristic of technology as a religious form is mediocrity, which is the chief characteristic of Western technology's Christian foundation. Mediocrity is what makes technology so appealing to the masses, because it makes everything possible; every hope, every desire, every wish, every longing can come within the reach of everyone. But what is possible dwindles down to what is feasible, and what is feasible are the small things. But the attainment of equality, which is so much a preoccupation of our time, ultimately gives the masses very little to be proud of, because they would have achieved merely equality with what everyone has, but they would not have what everyone has failed to attain. Consequently, technology as a religion lowers the prospects of man.

Since the masses embrace as well as encourage the ideal of mediocrity, they deny individuality, the promise of which is one of technology's great illusions. The greater the mass, the closer it approaches the lowest common denominator, and therefore, the more mediocre its character. This it cannot prevent. Technology cannot forestall the natural occurrence of mediocrity in a time when we have become so numerous. We possess just enough power to get halfway up the mountain. The existence of masses not only give its participants a false identity, but it also allows them to doubt their true selves, for the leveling process of mass culture is expressed as a denial of the self, as a form of self-hatred. And to acclaim the masses means eventually to obliterate the possibility for heroes. Heroes transcend the masses. Although they originate in the mass, they rise above the limitations imposed by mere numbers; they ascend to the top of the mountain, for a hero is worthy of admiration, an idol we emulate, because he is a person endowed with extraordinary qualities. Since there is nothing extraordinary about a multitude, that is hardly the place to find heroes. Of course, not even Christianity can be said to be a religion for heroes. It contains no hero worship because it strives to be mediocre, appealing to everyone. The closest Christianity comes to heroes are saints, people who achieve notoriety through mortification, shame and humiliation, through a consciousness posed against itself, none of which qualifies as truly heroic behavior.

It is also because we have no genuine gods who speak for us that we have no heroes; for wherever we have heroes capable of greatness we have gods with the same tendency--both are capable of performing mighty feats, just as each dares to defy the other. In place of heroes, our age has hero substitutes, that is, individuals superficially heroic, such as actors, athletes and popular musicians who are really surrogate heroes, whose popularity is sustained by their appeal to a mass audience, although their form of heroism is always peripheral. These modern heroes are idealized by the masses because they are believed capable of accomplishing wonders beyond the reach of the ordinary. Hence, they are worshipped. But as we said above, mediocrity and equality deny the hero. When genuine heroes exist, the situations which create them are far more discriminating, to say nothing about the idea that heroes themselves distinguish between alternatives. Yet the greatness of the hero rests on more than what he does. It rests on who he is, and who he is plays a part in what he does. To be at the right place at the right time is no more a qualification of a hero than luck is--no matter what else may be unnecessary to heroism, talent, that is, excellence, always is.

Hence, true heroes are particular, and all particularisms are anathema to technology because technology is the great equalizer. Like democracy, technology innately distrusts heroes because the latter conflict with the former's mandate for equalization. Because technology loathes differences between distinct entities, the lazy and marginally successful are paraded with the energetic and gifted so that not so much are the lowly raised as the exceptional debased. Too often the weeds are left to grow while the trees are chopped off at the top. In this connection, there is a reason why technology brings absolute decadence because it consists in the elimination of creative energy.[2] It deprives us of the source of originality, which has been severed from us, like a form of castration. Of course, if you are equal, you are interchangeable, and if you are interchangeable, you are expendable. Equality does not liberate the individual, but obliterates him. Nor does equality increase liberties because it shrinks the differences between the exceptional and the ordinary. If there were any laws of averages which regulated human affairs before the advent of the technological age, they were certainly reinforced after the dawn of it, for the kind of religion manifested by technology is truly for everyone. Even if technology fails ultimately to bring equality, at least it brings the demand for it.

Of course, technology does not propose final answers to the questions it asks. It is indeterminable. Traditional religions, on the other hand, propose determinate and obtainable ends for us. Whereas technology seeks to gain more knowledge within the infinity of knowledge, traditional religions cast a pall upon technological pursuits, particularly, when we consider the relationship of knowledge to technology and religion. Technology, now accompanied by science, is the pursuit of the knowable. Religion is the worship of the unknowable. The purpose of technology is to uncover and to probe the unknowable so that it may be knowable. The purpose of traditional religions is to accept the unknowable for what it is and to leave it concealed because God is hidden. We need divine intervention, that is, revelation, to make his presence known. We must seek him out to know him. Traditional religions ultimately are concerned with the unpredictable, the unintelligible, the denial of what is at hand, the affirmation of what is not, but concealment does not reveal the nature of God. It cannot say if he is good or evil, concerned about man or indifferent to him, approachable or aloof. All we know is that supposedly someone, somewhere believes that he should exist. As astutely delineated by Kierkegaard, we must make a leap of faith in order to believe in him. But technology has altered this concept. It has altered religion because it attempts to reduce speculation and doubt to fact, to replace wonder with absolute certainty. Technology attempts to erect a wall of scientific facts as a defense against revelation. It substitutes new myths for old ones.

Unfortunately, traditional religions become irrelevant by giving people meaningless banalities, by dispensing assurances of some distant and peaceful afterlife that would go on everlastingly *ad nauseam*. Oppressive and unrealistic, traditional religions become wastelands. They lack both leadership and practicality, and are far from understanding being. But, then again, traditional religions easily fall prey to technology's mechanization, which challenges and demythologizes religious mystery. Technology in particular and modernity in general have delatinized religion, making it as monolithic and as sterile as mechanized agriculture. Technology makes all religions useful, and thereby makes them useless. So we can see that the so-called death of God is either a liberation or a deprivation. If we can come up with something better, it is a liberation, but if we are to replace God with technology there will be little prospect for the perfection of being--because ever since technology developed into a system, the development of the self has been under increasing threat. Although Christianity in all its manifestations (prophetic, mystical and sacramental) is dead, the idea of God still lingers because we are always in need of him. We still seek him; but if we fail to come up with something better, our deprivation will make Christianity and all religions that much more attractive--a pity, since these offer us little consolation in an age so spiritually baseless.

In comparison to the past, the modern age repeatedly finds itself at a loss. We lose because although traditional religious values have slipped away, their unprovable prophecies still persist among us, giving cause for hesitation to many of us, but also because technology itself has a religious calling. No matter how we perceive the present situation, we live in uncertain times, for there is no one at the helm. Thus, the loss of spirituality, in the true and profound sense of the word, has been a great detriment to modern man. Such a loss has deprived us of the self-assurance that comes with a transcendent realm of being. We have been condemned to live in a world of objects, and like Adam expelled from Eden, we feel homeless in the world. Destitute and desperate, we make false idols in the absence of God and listen to false prophets. Thus, a new spirituality, hitherto undefined, is in great need but of little demand in the world, a spirituality that will both define the beauty of man and the wonder of creation, a spirituality that will go beyond any historical messiah, but which does not ignore man's humanity. A spirituality which indicates that, indeed, there was a Fall, not from God, but from ourselves. So here is the principal accusation that we can make against our age: just as Christianity survived and prospered in the ancient world primarily because of the decline of freedom, technology develops in the modern world out of the same perversion of what is rightfully meaningful.

Just as civilization developed in spite of religion, though we hear this truth denied by all religionists, so too we must perfect ourselves in spite of technology. Before we witness what some believe to be the inevitable triumph of a dispassionate technology, we must promote an age that gives itself over to a spiritual rebirth, to a new age utilizing technology, not utilized by it, to a time when man will regain his power, his purpose, his very presence. We must speak with a voice prophesying a new age. Such an age must place man at the center, and not utilize him as a means. It must be an age given over to sharing and caring, even if not necessarily free from want. It must be an age satiated with life, fulfilling us to the end of our days, making us strong in the face of adversity, helping us bear with stoic resolution our sufferings and misfortunes, allowing us to face death honestly because we believe in life. Technology does not concern itself with man per se but only with man as a means in order to procure ways of doing things, that is, with devising techniques for the execution of actions. Although Christianity withdraws from communication with the world by renouncing it whereas technology supplants it, a new spirituality must not allow us in the least to transcend the world, that is, it must tie us to it. It must not leave the world behind. Technology rationalizes the world; it makes it comprehensible, and therefore puts itself in a position to exploit it. It teaches us about things, but nothing about ourselves. It deals with the world constituted by us, but ignores us as the agent of the world. It makes us look externally for what can only be found internally. It purports to make us tenacious, but really makes us weak. It is small wonder why this is so, since we are so much removed from ourselves. But in order to understand the struggles of existence and the limits imposed by it, we must not forget ourselves when gazing at the stars.

Notes

1. Cf. Locke, *An Essay Concerning Human Understanding*, ed. Alexander C. Fraser, 2 vols. (Oxford, 1894), bk. II, ch. 28, § 5.

2. Ortega y Gasset, *The Revolt of the Masses*, p. 43.

Chapter 8

A Shortfall in Knowledge: Ignorance and
the Proliferation of Information

Knowledge exists either because we uncover it through discovery or we create it. If it is discovered, then it existed before we uncovered it, and it transcends us. If it is our creation, then it did not exist before us, and it is anthropomorphic. Obviously, how we perceive knowledge determines how we define it, but knowledge can only be knowledgeable when it is knowable. Most of us agree that knowledge is created, not discovered; that knowledge cannot exist without man, since knowledge is linked to subjectivity. Perhaps, we do best to avoid the use of the term "truth" entirely when we intend to signify anything other than the being of being (*aletheia*). We should speak only of what cannot be avoided, and that is knowledge. We know that there is knowledge, that it is knowable, and that man is the knower. Ultimately, truth is the disclosing of what is, cloaked by man in a world created solely by himself. Everything else is nonexistent, and what is nonexistent is not.

Technology constitutes part of the world created by man and comprises part of the knowledge knowable by him. Thus, we can agree to a limited extent that *techne* signifies more than art or skill because it also signifies knowledge skillfully derived.[1] But technology's "knowledge" is very much occupied by limited and insular bits of data, defined and cut off by time and circumstance, which we must differentiate from what is commonly called true knowledge. These insular bits of data are merely informational, and they must be judged as knowledge of a lower level, for they are collected and gathered up while pursuing something else. Thus we must distinguish information from knowledge. Information is a type of knowledge that is made to order, rapidly prepared and derived from well-defined and easily manageable tasks. It is founded upon the mastery of techniques, which in themselves constitute a certain type of knowledge, but not one we ordinarily equate with it. Under the most favorable circumstances, technology's conception of knowledge is limited to what is analytically verifiable, to quantifiable fragments of reality, to bits and pieces of objectivity. Thus, ultimately, it stands outside and apart from man, for it is foreign both to objective uncertainty and to subjectivity, for even the laws of relativity depict an unstable world, at least from the eyes of the observer.

In point of fact, to know means to acquire. To inform means to impart the knowledge that has been acquired. Knowledge signifies recognition of something; information is merely a form of news, of events and happenings, but more ordinary than extraordinary. Although knowledge is an end in itself, information is a means to this end at best. And the means that imparts information in our world is technology. Preoccupied primarily with information, technology is hardly in the least concerned with knowledge as it has been defined above, because whatever the end of knowledge might be, technology can be no more than the means of its conveyance. Because the rapid and seemingly endless proliferation of information has led to the fragmentation of learning, more and more areas of information have resulted in a greater ignorance of all of reality. Although we know more today than we did yesterday, we also know these things from a more limited point of view, as from the perspective of a microscope, which is a most tubular, that is, a most restricting way of looking at the world, for we are brought up closer and closer to the subject of our intent, while ignoring everything else. What results is the specialization of information, augmented by endless proliferation, which itself is a cause of specialization. Since specialization means learning more about less and less, we enter a vicious circle from whose embrace it is difficult to escape, a vortex of complex structures and rigid procedures which swirls us deeper and deeper the more we struggle to make sense of it.

The rapid expansion of information has widened the gap between its acquisition and its assimilation. This hiatus poses a serious threat to the usefulness of newly-acquired information, and has given rise to the creation of two different types of technicians, both of whom are adversaries: superficial generalists, who have nothing to teach, and narrow-minded specialists, who will teach us nothing. To acquire more and more information about a smaller and smaller aspect of reality is not learning, but merely increase in information. If it were true learning, it would be accompanied by humility, since the more we learn, the greater tends to be our feeling of what else remains; but such humility is strangely absent. Moreover, in the past, learning gave a feeling of certainty because even though we knew less we thought we knew more; today, we have the same feeling of certainty because we know more quantitatively, but in actuality it is far less than we think. Indeed, scientists and scholars have the tendency to become buried by their own learning, to be so inundated by erudition as to be incapable of distinguishing between genuinely meaningful knowledge, expressed in the form of existential know-how, and scientific data that exhibits itself as information having useful or pragmatic applications. Although conditional, the purpose of knowledge is to make us transcendent; but far too often we encounter disjunctive information, facts that are unconnected and unrelated. Likewise, there is a major difference between the past and history--the past is what we have salvaged in raw evidential materials, but only their interpretation constitutes history. A fact may remain as it is; only its interpretation changes. The best that we can say about a fact is that its interpretation determines its relevance, although we far too often resort to facts before we have had time to interpret them. (Indeed, a fact is the recognition that a thing exists or has existed in the world. And the world, contrary to Wittgenstein, is made up of things--which may then be designated as facts--not facts initially. A fact is merely a judgment upon a thing. Since things lie on the periphery of man, we cannot say that only when we have come to know all things will we come to know him.)

The modern age holds to what may be an emasculated view: that new information has a greater value than careful rethinking of old information, that the latest advancement in learning, no matter where it leads, is ultimately beneficial to man. But more tellingly, the very word "information" now takes precedence over "knowledge," for information signifies the latest available manifestation, the most current presentation of currentness. In general, information means representation, a depiction of the object of one's intent, a conception of an idea, all of which are divorced from feeling because information is cold, hard and unmoving. But knowledge is a far more fundamental description because it is a confession, a laying bare of vulnerability, a declaration of faith, an adherence to driving forces and beating hearts. Knowledge is attuned to commitment, that is, to responsibility, not to technological or scientific precision.

Modernity encourages us to gleefully disregard everything tested by time, to walk away from the wisdom of the past. Although there are also limits to what we can know, it is often difficult to distinguish between interpretation and use. There are even limits to what we can know of the self, our greatest enigma, which remains indefinable and unconditional, forever open and unsolvable. Not only do we assign a greater value to new information, we can no longer use it to explain and elucidate human experience. We use it instead to uncover and analyze what lies behind it, and technology is the means how this is done, which is able only to tackle limited and well-defined tasks because it is little more than a mode of discovery. Hence, technology is never an object of knowledge, but a means to information. Of course, there has also been a decline in the quality of information in comparison to its quantity, to say nothing of the spread of scientific information diminishing the value of humanistic knowledge. So we can now elucidate the differences between an atom and a molecule, an insect and an arachnid, an epiphyte and a parasite, but we remain unenlightened about the differences between liberty and freedom, essence and existence, being and nothingness. This dichotomy between scientific information and humanistic knowledge is one reason why the present age is in such great need. Then again, perhaps truly transcendental knowledge is an unattainable dream; perhaps it is reserved solely for the gods.

Science is incapable of asking all questions. Hence, it cannot give answers to all questions. And the answers science is capable of giving are incomplete and fragmentary, since science is the search into the infinity of knowing. As a result, the humanities become all the more important because they give us a heightened sense of awareness that allows us to ask the most meaningful questions, not questions asked by a mind accustomed to a limited and limiting mentality, but questions posed before the plenitude of being by a free and open being infinitely inferior. Needless to say, true science is an end in itself; it is simply the pursuit of knowledge for its own sake. But when science becomes coupled with technology, the scientist with the technician or engineer, science serves another end, the end of utility. Yet science is not free from the abuses of technology because technology is merely the manner by which science is pursued. Although it may be said that science does not ask valuative questions but deals only with facts, this statement is misleading because the statement itself presupposes a value, the value that science somehow is divorced from all values.

Modern science concerns itself with understanding the workings of the universe, but it cannot ask valuative questions about who is man. If science could ask and answer these questions, then the science that evolved would be different from existing science. Thus, whatever philosophic truth is, it certainly is not scientific truth. Nevertheless, science has radically changed the way we look at the world. Since Descartes, science and the technological age which supports it presuppose that truth is objective and that the passions are the slave of reason. Cartesianism superimposes mathematical certainty upon the arbitrariness of the natural world, but the quest for certainty severs man from the world, making him a solipsistic cynic. When technology and its method of acquiring information by means of science apply in any way to man, they cannot help but to distort his quality, to oversimplify him, because they reduce him to a collection of manipulable and analyzable data. Although science tells us that ultimately truth is an external manifestation of being, being tells us that truth is an internal manifestation of being itself. Hence, the conclusions of science, which are presumed to rest upon a coherent theory of truth, themselves rest upon subjective evaluation.

At one time, the pursuit of science was thought to be truly innocent and totally good, devoid of selfishness and pettiness. We now know otherwise. Although technology increases possibilities on the one hand, it does not render all these possibilities desirable on the other. Therefore, we must modify the excessive optimism put forth by Bacon, whose *Novum Organum* happily advocates the tyranny of science, but whose scientific precision proceeds at the expense of the humanities. Science and all mathematical disciplines, such as logical positivism, postulate that only they are precise, that only they are exact in their assessment of reality because only they are analytically verifiable. In actuality, however, all measurements of reality are subjective, even when measured by an objective device--a device conceived and built with information acquired by subjective beings--and the sophisticated exactitude of science always retains imprecisions because it is humanly derived. Science is precise within the perimeters of science, given what forms of measurement it must work with, but outside of the reality it studies, science is imprecise. Indeed, the absence of universality is one of the failings of science, since if it cannot analyze and quantify every aspect of reality, it labels exact whatever it can analyze and inexact whatever it cannot. Therefore, human reality, which is beyond the scrutiny of science, is considered to be imprecise.

These views tells us a great deal about science in our day, that although the scientific method is simply one approach to reality, it is the best approach so far devised for understanding natural phenomena. But beyond facts, which constitute only physical reality, science tells us nothing. Human reality, that is, existence, will never be scientific because man will forever remain elusive to the attempts to categorize him. Hence, the basis of man's ultimate reality will never be in accord with science, nor will science be able to rewrite and reconstruct the humanities in its jargon and keep them as such. Even history is scientific only to a limited degree. Our notion of scientific truth is so abstract that it approaches an impartial truth, a truth so alien to man, as Husserl has said, as to be devoid of him.[2] Slowly but consistently, we have removed ourselves from the larger concept of truth; truth is no longer *aletheia*, the being of being, but conformity of perception to beings. Subjectivity now lies buried beneath the tonnage of objectivity; being suffers as a result of beings. Consequently, there is an inverse ratio between the knowledge of the self on the one hand and science on the other. The more the world becomes subject to the scientific method, the less we know about ourselves. Although we have the power to shape the future, this ability does not mean that we can truly understand ourselves or the world. We will always act in and through ignorance. The modern predicament aggravates this condition for it not only evokes uncertainty, it also lacks confidence.

By itself, rapid proliferation of information regardless of its origin merely buries it, and the older the information, the deeper it is buried. In fact, the early obsolescence of information and its continual replacement constitute one reason why technical information is followed quickly by oblivion. The rapid proliferation of information also embellishes a lack of understanding in its entirety; it makes information disjointed and promotes a loss in the sense of direction. Thus, too much information becomes unmanageable. Likewise, information is irrelevant if it remains unconnected, if it is unsupported by other related information. It must be connected in one's head and to one's culture in order to be meaningful. The proliferation of information vulgarizes significant achievements and turns them into so much rhetoric, to say nothing of the fact that each succeeding generation that acquires more information is less capable of digesting it than its predecessors.

The same holds true even for mass media. Unless we can relate personally to an event in the news, we will be indifferent to it, even to descriptions of great suffering and unmitigated tragedy. In our technological age, much information portrayed as news is fragmented and disjointed, enlarging the indifference that is already man's plight. The individual rarely relates to it because it has nothing to say to him personally, nor is it of historical interest. In fact, too much information, most of which is trivial by nature, is no better than too little information. And the accumulation and preservation of this marginally useful data, which is what we have become accustomed to, is data that in any other age would have been prudently forgotten. Unavoidably, the enormous mass of this information hinders the flow of its assimilation. We no longer have language, but information systems. We no longer have facts neatly organized and meaningful to the well-being of the individual, but databases manageable by computer. Although too much information leads to misinformation, far more often it leads to indecision. Since we lacked the ability to integrate all information into a meaningful whole even before the advent of our own age, we are hardly in a position to be able to do so now. Now, a consistent and unified vision of the universe has been rendered impractical and impossible. Rather than integrating all information, we have achieved instead only integrated ignorance, for there is a fundamental difference between information that merely supplements us and information that sublimates us. We lack not only the capacity for integrating information but the humility as well.

Of course, increased information leads to increased specialization. Since specialization means limitation to a discrete set of concepts requiring that small fragments of information be avidly pursued, it aids the spread of information, but hinders its understanding. Increasingly irrelevant the more specialized it becomes, specialization suffers proportionately to the extent of its limitations. Hence, specialization is manifested by the smallest and most incomplete aspects of learning, by breaking information into smaller and smaller fragments. Because of specialization, we have become an age characterized by the absence of great thinkers, who have the ability to transcend the limits set by their disciplines. The dangers of specialization have eroded the need for creativity and have replaced it with erudition. In

A Shortfall in Knowledge

fact, there is so much erudition, so much scholarship in the modern world that communication among scholars is difficult. Even the simple task of keeping abreast of one's own discipline is a burden. The very mass of information is so overwhelming that it hinders understanding. Perhaps the increasingly technical nature of all scholarly writing is indicative of the vast areas of ignorance scholars manifest outside of their major discipline. A specialist studies but one possibility, examines only one part of reality, works toward only one horizon. One specialty rarely influences another, because each specialization digs deeper into the dark strata below, tunneling its own way through a hardened mass. Rarely, does it rise to the surface, exposing for all what it acquired by its subterranean labors.

Because scholars, indeed professionals and even technicians, are highly specialized, they become incapable of communicating their specialty, except on the most mundane level, with anyone unless they are like themselves. All information burdened with specialization becomes arcane and secretive, meaningful to a few, mysterious to most. Specialization produces an esoteric, inaccessible and sterile elite, indifferent to other specialists--an endeavor by professionals for professionals. Even modern art is now specialized; it is art for artists only. Nevertheless, these specialists themselves are dependent upon others for information which does not directly concern them. Each is a cog in the wheel, whose ultimate purpose eludes the other. A specialist in any discipline is simply a technician with a limited scope working in a dimly lit laboratory surrounded by perpetual darkness. Or one might say he is a princeling who walls himself up within a great tower, cut off from the outside world by portcullis, parapet and moat.

Specialization comes into being as a direct result of technology, because it is created to deal with problems posed or functions generated by technology; it is not, as many think, the direct result of science. Unmistakably, specialization never existed, nor had any reason to exist, before technology played such an appreciable part in the acquisition of information. But once established, specialization and the specialized techniques which accompany it become limited to rigid assumptions and expectations, which help to define their peculiarity. Any increase in specialization ultimately leads to more organization, which in turn leads to more specialization, since the more we specialize the more we need new ways of dealing with newly acquired information. Because technology renders things complex, its results are probable only within well-defined limits, which are both delicate and unpredictable. Although we have become more organized, that is, more structured, we have become less capable of assimilating specialized information. But more to the point, specialization exists because we distrust what we cannot measure, since our age loves to categorize things. The immeasurable is beyond our control, thereby making us uncomfortable. And, of course, the measurable is the exploitable, which fits well with technology's view that we are to show no pity for what we can exploit. Hence, we strive to render all of reality measurable so that we need feel no emotion in an emotionless age.

Then how can we comprehend the truth that we ourselves embody? How else can we answer when we know that man rarely receives the attention that is owed him, and rarely does an age produce a philosophy that regards him to be the subject of its design. It is customary for man to be last, the last subject we study, the final horizon we attain. But information is acquired by us for our benefit through our views of perception. The whole universe is anthropomorphic. All information is tainted by our hands. Then who can deny that truth is subjectivity? How is it possible to rid man of man? We can only meet with laughter the view that truth, in order to be truthful, must be cloaked in "p and q" statements. Just as science becomes scientific through inclusion, philosophical truth becomes scientific through exclusion, and that is its downfall. Technology and the techniques given over to it work day and night to destroy the homocentric view of the world. Technology conspires that all manner of things may be generated only by itself, that it exists totally in the absence of man, that in fact it is devoid of him. Because the modern age is characterized by abandonment, because it gives itself over so voluntarily to technology, it is an age which symbolizes the antithesis of man. Such an age becomes incapable of building anything in man's image because we have betrayed the spirit of our being, denied its existential underpinning, and removed it from being itself. We have forsaken affinity with our own humanity. Thus, we live in an age of disillusionment, in a time of retreat producing nothing except more of the same.

Specialization also signifies the dehumanization of knowledge because it not only deprives it of its intrinsically human relationships, it also categorically denies that man is the object of its purpose. Unless knowledge serves man, it serves no one. Indeed, the crisis of humanism is the denial of man. Hence, we have evolved--or devolved--from a knowledge that is helpful to us to one that is harmful, in fact, to knowledge that is fatal. We have devised weapons that would destroy more than an army here and there. We have devised them to obliterate our cultures, to annihilate the past, to change the very face of the earth. We have made everything possible. We even make ourselves in vitro. Thus, knowledge has succumbed to the mechanization of our time. In point of fact, psychology has been replaced by behaviorism, a discipline whose sole purpose is the control of man, which is truly an obsession of our time. Even philosophy has been reduced to logical positivism, which characterizes the debasement of philosophy, and is an example of how philosophy is perverted into a science. Yet science is the least capable of solving problems because it already perceives the world as complete. It views the world as a closed system. It makes the world disconnected. The old, murky questions of life have been abandoned for the analytical questions of elementary mathematics. The questions of metaphysics, of ethics, of political thought are now considered inferior to any question which can more easily be answered. It is true that technology can deal with analytical questions, which is why they are technical, but it is incapable of dealing with metaphysically based questions, and so it ignores them. Technology so simplifies philosophical questions that they either are considered meaningless (when they are equated into mathematical or linguistic terms) or they are substituted for pseudo-questions, whose depth of perception is but a fraction of the profundity of philosophy's great questions. There are even those who consider metaphysics a nuisance, who perceive these great questions as hindrances to progress.

The more we remove ourselves from being, the greater is the distance from truth. Wherever there is truth, there also must be being. Thus, there is no such thing as unbiased truth. To assume that technology is flawless because we do not enter into its unbiased conclusions ignores the fact that it is man who furnishes information, and all information is subjective and incomplete. In many respects, technology has engineered a great injustice because it ignores man's subjectivity. Ultimately, technology is indifferent to us; it is like a stone--cold, hard, uncaring, yet it adamantly denies that there are other ways to knowledge. The metaphysical danger of technology is that it and it alone may one day be believed to embody truth, casting everything else into an abyss of falsity and half-truths. Hence, technology becomes the means of prevarication, and to it belongs the route of self-deception. Its conceptions of truth are cloaked under the supposition that it has no axe to grind, that its alleged neutrality should offend no one, making a friend of everyone who believes in progress. This so-called neutrality, abetted by the language in which technological knowledge is cloaked, deliberately attempts to obscure all issues, particularly all moral issues, as if morality no longer is an issue. Of course, this cannot be done; nevertheless, technology purports to be able to do this, to put itself above morality. Since value judgments are normative, they cannot be divorced from knowledge. The value judgments of technology, therefore, are supposedly based upon data that indicate their likely effects upon us and nature. Value and knowledge are not diametrically opposed to one another. A value expresses a choice, including the choices manifested by technology. But technology conceals its value judgments with misleading and confusing reasoning slanted in its own favor, and by doing so it ignores or even willfully sidesteps its true obligation.

Because technical information is accessible to anyone trained in its methods, it is information without quality. It is information, raw, pure and simple. This nakedness is reason enough why technology cannot be monopolized by anyone for very long as long as it remains overt. Technical information, like technology itself, is contagious, communicable by contact, diffuse. Thus, it spreads from one culture to another, from one group to another, from one person to another. Attempts, for example, by England to prohibit the exportation of any information on machine technology failed to keep her American colonies economically subservient. Similarly, the Ottomans, previously ignorant of artillery, learned how to use it from the West and with it battered down the walls of Constantinople. Greek fire, the atomic bomb, or laser technology--they all suffer the same fate. Even when legally protected from infringement and cloaked with national prohibitions and religious sanctions, technology's universality allows it to be the object of theft and piracy. Yet, ironically, technology may be lost to future generations, because it is perishable. The technical information upon which our age takes immense pride becomes obsolete by a natural, evolutionary process. But the greatest irony to our accumulated mass of technical information would occur if future generations deliberately ignored our highly technical books in order to pore over books on horticulture, pottery, carpentry and basket weaving, which in themselves may be technical, but hardly in the light of astrophysics, genetics and linguistics.

Training in the humanities, which now appears to be outdated, is a superior form of education because it prepares the student in all his faculties, because it gives preparation in the sciences of the spirit (*Geisteswissenschaften*). A liberal arts education contributes to the cultivation of a critical, well-informed citizenry, capable of participating in society. Although being better informed than those in the past does not necessarily fare us any better when understanding the world in which we live, one may say that an intelligent man without an education is still superior to a man poorly educated, but a classically educated man is superior to both. A technical education, on the other hand, too quickly becomes obsolete; nevertheless, education has always had a vocational side, since acquiring skills for employment is preferred to non-vocational instruction. Indeed, knowing how to read Latin furthers one's education, but generally does not help one secure employment; and on that account it is no longer valued, and seldom taught. Education as a result no longer confers wisdom because it has lost the power to distinguish quality, because it has no sense of direction.

Thus, an age conditioned by technology produces two equally inferior, yet different types of people--the superficially and technically educated as one type and the ignorant as the other, that is, the few and the many. On the other hand, technology produces people who are extremely educated in isolated pockets of learning but who are poorly educated in all the others, a situation that is particularly aggravated by the presence of specialized information so characteristic of technology. These individuals are Whitehead's "minds in a groove,"[3] who are accustomed to bursting in with such force and with such assurance whenever they open their mouths that they believe only they speak "the truth." Those who are well-educated in some small area are generally impenetrable in all the others because specialization closes off the world. But in order for these few individuals to work in the clouds, many must toil beneath the earth, for original research pursued on the threshold of the unknown would be impossible without armies of minions who sustain society by performing the most rudimentary tasks. The masses support the few; and like a pyramid, the higher one goes, the less room there is at the top.

But the silence of the masses has been broken by the great increase in information. They have been included, whether they like it or not, in the daily barrage of data produced by technicians and specialists in all manner of learning. We now hear them chatting about economics or automotive engineering or ecology. As a result, the masses have an opinion about everything, especially things about which they know little or nothing at all. They do not know that their learning is limited, that information is always conditioned by circumstances which require the continuous need for enlightenment and so for them there is little hope for improvement. Although culturally eclectic, they are intellectually dead. As philistines, they think and act as if they are the epitome of refinement while they enshroud the possibility of truth with clouds of ignorance.

Notes

1. Heidegger, "The Origin of the Work of Art," in *Poetry, Language, Thought*, trans. Hofstadter, pp. 59-60.

2. Husserl, *The Crisis of European Sciences and Transcendental Phenomenology. An Introduction to Phenomenological Philosophy*, trans. David Carr (Evanston, 1970), pp. 124-125.

3. Whitehead, *Science and the Modern World* (New York, 1925), pp. 127-128.

Chapter 9

A Disparaging Condition: Challenges to the Self

The presence of technology is a threat to the development of the self because it challenges the need for freedom. Technology gives us the feeling that we no longer have to be authentic in order to act authentically. Regardless of other factors already discussed, the world has been well prepared for a technological age, but more than the lack of rights, technology means domination. It means the submission of individuals to a master's caprice and tempts us always to say yes because like bondsmen we learn very readily, through threats and other persuasions, never to say no. Freedom, if it is anything at all, is a negation. Because technology limits alternatives to a paltry few, when in fact they are plentiful, its sphere of influence is always smaller than that sphere which remains, yet technology strives to diminish the impact of individuals in the grand scheme of things.

And what is the individual but the self. There are as many selves as there are individuals. The self is the manifestation of a subject, a person, a unity of consciousness and will, freedom and choice, action and deliberation. The self denotes the concrete being, made whole through self-expression. The more indivisible the self, the more individual, that is, the more inseparable the self is from itself, the more undivided is its being. The self is not an inclination, but an existent. It is an entity, not a tendency, and certainly not a disengaged observer, nor merely Hume's collection of perceptions.[1] There is nothing disengaging about the self, since to be a self means to be engaged both in one's own being and in the manifestation of that being in the world. Even when suffering from schizophrenia, there is still a self, although it is disoriented and bewildered. A self has been defined quite accurately by Boethius as that being which is undivided in itself and divided from anything else, to which we must add that although the self is separate from the world, it is still tied to it.

Because the self is a unity by its very existence, it cannot be described as embodied, but neither is it unembodied. The self is neither all body nor all spirit. Although we act through our bodies, we are not our bodies. The self is the totality of a person; it is everything a human being is. And for this reason, the self is not a fullness or a plenitude nor can it be filled up. When the self comes into the world, it already is a totality. Nevertheless, the self is continually enlarged by new experience. Although finite--how can one be other than oneself?--its character and development are always subject to modification. This is how the self develops as a person, how it becomes aware of itself, its world, its past. The self must change to survive; it must evolve. Above all, a self must be willed in order to become a self; it must realize itself through its own efforts. The self is made by choosing itself, made patent through latent possibilities. Since we individually determine the nature of being, to will being is to will oneself, since being means everything that a person is. The unique quality of existence, therefore, is selfhood, the irreplaceable condition of being. We have a self, in part, because we are conscious of ourselves, but we also have a self because we are free, because what the self is depends upon what it becomes. Since we have possibilities still unrecognized in our future actions, we are not determinative or definitive. Thus, one should reiterate what has been said so many times before that one is not what he is, but is what he is not (yet).[2]

In addition to free choice, the self is also identified through the passage of time. Every self, that is, all of existence, is conditioned by a measurable duration, by a reckoning of an indefinite but finite continuum, by an inescapable emptiness waiting to be consumed. Time is a measurement of an eternal present, of a present still forming out of chaos, emerging from the mist of the future, but once consumed, rendered irreversible and unalterable, forever frozen and crystallized into the past. Time must have movement to give meaning to the self, not perceived as standing still. For each individual, as for everything, birth begins the process and death ends it. Thus, the self can pass away completely, that is, it is perishable because it runs out of time. But death does not mean that we have caught up with our past; rather, it means that the past has caught up with us, and thereby holds us forever in its grip, forever entombed like a mummy in a pyramid of timelessness.

No matter what terrible experiences the self may encounter, it is safe from violation. It is impregnable. The self enters the world as a totality and leaves the world in the same condition, although enlarged by experience. The self may be cajoled, threatened, tricked or duped, but it remains inviolable. There is never any actual loss of the self because the self cannot be other than it is. It cannot exist in any other form than in its existent. The so-called loss of the self and flight from the self are misnomers. Nothing is lost from the self when one has a self, nor can one flee from the self. Never can a void or emptiness appear because of lack of selfhood. As already stated, mental anguish and psychoses put the self through much suffering, but as painful as these experiences might be, the self remains intact, albeit strained and abnormal. Although we can be disoriented with ourselves, since we can lose the sense of the self or even be estranged from ourselves, there is no way that we can leave our individual self behind. As incomprehensible as psychoses may seem to us, we never can leave our individual self behind. Like a shadow, the self goes wherever we go.

The self implies separation from those who are not oneself. To experience oneself as an identity is to experience a dichotomy between the self and all others. Not being able to look upon oneself as apart from others is to call into jeopardy one's own selfhood, since to stand against "the other" manifests the emergence of the individual into a self. Of course, it is not man in general who seeks a self, but each individual man. We do not ask for a universal self, but for our own selves, although the existence of the self does not precede the discovery of the self as an existent. It does not precede the authentication of the self by the self nor the alienation of the self from everything else. The self exists only by participating in the world because that is how the self is recognized to be a self; the world is a creation of man just as man is a creation of the world, and through man the world is perceived as his creation. Thus, the self exists because it is called into play by the world, and the more we make our individual selves ourselves the more authentic our being. Since an individual self must be aware of who he is, this awareness also implies who he is not. Thus, how we relate to others is an essential part of being, but how we do not relate is also. Furthermore, the past also plays a part because the self must be aware of his own past, of what he did and did not, of who he was and was not. Not surprisingly, we are different from the way we were in the past; but we also carry the past with us today, since it has made us who we are today--as we will carry today with us tomorrow. Thus, a self is defined by both time and choice, by both the past and action, by both temporality and freedom.

Nevertheless, we can always deceive ourselves, since we can live a denial. We can block out the past when we find it disagreeable or unfulfilling. We can deny the past when we find it to be radically different from the past of others. Because the self penetrates the conditions into which it is immersed simply by the presence of its being, the very presence of technology as a systematic methodology tends to jeopardize selfhood, particularly because the self is more apt to act without self-reflection. Threats to the security of the self are extenuated and intensified by technology. Since self-reflection means to attempt to find oneself in existence, anything which hinders this endeavor is a threat to the self. But self-reflection has meaning only to prepare us for interaction with the world. Self-reflection is to aid us, not consume us. Because many of our actions can be unconscious, it is imperative that the world in all its diverse forms, including technology, be filtered out by us when we need to understand ourselves. Not that we should say no to the world (how could we do otherwise?), but that we should say no to an automatic, unthinking response to technology's eternal presence in the world. Otherwise, we may never allow ourselves the opportunity to do so because we will never be alone with ourselves. Since technology is possessed of systems and rationalities already devised and set in place, which in turn are augmented by instantaneous gratifications and self-deceptions, we are at a great risk. But technology posits a threat in other ways because it gives us a course of evasion. It gives us an excuse when we wish to live inauthentically.

Surprisingly as it may seem, this threat is something that the self cannot avoid because it is not possible for it to do otherwise. There is a self because there is the world, and by the world we mean everything that is external to the self. The self is tied to the condition of the world just as the world is tied to the condition of the self. The world is given birth by the self just as the self is created in relationship to the world. One cannot exist without the other. The self cannot avoid its immersion in the world, no matter what type of world it may be, or any aspect thereof because the self and the world are inexorably tied together. This juxtaposition is most notable in the case of technology because, unlike all previous ages, none of us can escape to an isolated hermitage and be free from technology's influence. We live in a time, contrary to all previous ages, when technology's driving force penetrates everything, when there is no longer any possibility of not living without technology, of not living without immersion in its world view, without being soaked deep in its conception of how to think, how to act, how to be. We can run from technology, but we cannot hide. And by technology's influence we mean its methodology, its excessive rationality, its schematic systematization, its pervasiveness.

Not only does technology jeopardize self-expression, but other factors also augment this condition. Because there is little real change in an age founded upon technology, there is little opportunity to be truly active, and therefore, less of a chance to be genuinely a self. There must be real meaningful change in order for the self to grow. This is true despite the possible refusal of the self to identify with itself, despite the unwillingness of the self to recognize that it is apart from all other selves. And because we may be absorbed into a world of techniques and rationalities created by technology's efficiency, we can easily allow the world to entice us to follow its dogma, that is, its manners and mores, its prescribed ways of thought and action. So our lot in life becomes harder, that is, more of a threat to the self, the easier technology makes our lives. Although technology creates needs, it is also created by them, and these needs go a long way to making all of us pursuers of mere externals, many of which are not really needs at all, as we said above, but wants disguised as needs. Although needs are limited and essential, wants are unlimited and non-essential. The so-called needs of technology are not just extrinsic to the self, but extraneous to it. And as we make the world more dependent upon technology, we become tied to it more and more. We wish to become part of it, and therefore retain fewer opportunities to be true to ourselves. Distrustful of ourselves, we feel compelled to listen to what technology can do, that is, we put our trust into technology's mechanization while simultaneously distrusting freedom's openness to being. For we have been colonized, induced by technology to strive after things foreign to ourselves. Thus, we attempt to master whatever is external to ourselves.

A technological age is characterized by the anxiety of having no place, of belonging nowhere, of being immersed in a void without form or direction. But it is no irony that it is also characterized by its opposite, by the anxiety of being imprisoned within a hectic and restless life without hope or purpose, that one has become part of a thing without willing or desiring to be so consumed. Although bombarded by indecision and lack of cohesion, we encounter and are still responsible for our own risks, disappointments and limitations. Human existence is always particular because it deals with the finitude of human reality. If we struggle to achieve the perfection of our being no matter what else happens to us, we will never fail existentially, but we too often would rather not do this. We would rather have others decide for us, or lacking that, we would rather have technology make us comfortable so that when we choose, it will be as painless as possible.

In point of fact, technology has complicated life's uncertainty and indecision because although technology is an ally to having as well as to doing, it is an enemy to being. And when it comes to being, we should reflect upon the thinkers of the past. If their understanding appears to be greater, their perfection deeper, their tranquillity more congruous, it must be attributed to their close touch with understanding the being of human beings and to the circumstance that their knowledge of man was not removed from the reality it represented. Although attuned to being, the past also possessed less diversion, less dissipation, less distraction. Being was not a stranger to itself. But our preoccupation with technology and blind disregard of being goes a long way to explain why we have failed to acquire a better understanding of ourselves, why of all the ages, it is our age, despite the modern efforts of ontology and psychology, that suffers from such a poor knowledge of man.

But other factors are at work against the fulfillment of the self. We become increasingly regulated and restricted by technology's influence because technology reinforces conformity and challenges any deviation from the norm. And it achieves this by imprisoning us, by confining us, by walling us up as in a fortress. In a sense, it buries us. Technology covers up existence; it undermines the self. Since technology serves as an obstacle to the perfection of being, it is nonsensical to presume that technology is ontologically derived.

Because of the broad range of its adaptations, variable to time and place, it seems more feasible to say that technology emanates from being's free expression in the world, made manifest by morality and history, that is, technology is culturally conditioned. Despite its origin, the modern age suffers doubly for the use of technology because not only does technology obstruct being in general no matter what age one lives in, an age which uses technology as the basis of its culture becomes engulfed that much more. An ever-increasing accumulation of techniques, products, devices and machines buries being. Thus, the more we perfect and utilize technology, the more the self is threatened and endangered.

One assumption made of technology is that it allows us to think about ourselves, presumably because it gives us more leisure time for reflection; but it does not. Technology fails because we become dominated by its very presence, by its devices and techniques, by the complexities of its rationality and the convolutions of its methodology. Technology cannot help but drive a wedge between us and self-awareness, between us and that relational phenomenon which is grounded in inwardness, that is, in the awareness of the individual of himself, of a kind of self-directedness, a reflection of the self to the self. Until we make a conscious effort to remove ourselves from technology's driving forces, it will continue to reduce our prospects of liberation. Like an object which has fallen into a river torrent, we are violently swept away because technology moves forcefully in only one direction. Since we invest so much effort in technology, we undeniably project high aspirations upon it, and the greater technology's power, the more we project ourselves.

It is well-known that a technological age is characterized by the embodiment of impersonality because it compresses strangers with no common cultural or ethnic similarities into a disunifying whole. And after we are thrown together, we are encouraged to conform, to be like others--but in fact to be a nonentity, to be as Kierkegaard has expressed it, "a number, a mass man."[3] But technology also threatens the desire of each individual to occupy the principal place in the life of another; it ignores the necessity of the individual to be in the only significant place in the life of a loved one. And apropos of impersonality, technology puts pressure upon us to accept moral views and political positions which are not our own, that is, attitudes and beliefs which we would rather not support, as when we litigate for the rights of those whose views are contradictory to our own. Internality, which deals with uniqueness, with the inherent ground of being, contrasts sharply with technology's externality, which concerns how we conform to an image not of our own making, to an estrangement of our true self. Therefore, a technological age challenges internality because our point of reference becomes external to ourselves.

Of course, in order to maintain selfhood, we must be faithful to more than ourselves. We must also be faithful to the past and to our blood, that is, we must be aware of those who have gone before us, of those who have made the present, of those who have made us. It is self-defeating to assume an identity which is not our own. Like an adopted child, we must know our parentage. If we do not know it, we may be forced to assume a new one, but one which has been falsified. Although technology rarely disturbs the intimate relationship of the family, it does challenge one's membership in an ethnic or racial group on the one hand and the actual validity of such a group on the other. Technology's inducement to abandon the past also means our own past, not only the distant past of centuries gone by, but also the past of our immediate ancestors. The very denial of this heritage is exemplified in technology's pluralism, in its monolithic cultural wasteland.

Although influenced by things external to it, the value of the self is ultimately conditioned by its existential qualities and by its being in the world. Otherwise, the self would have no meaning. It would not be distinguishable from externality. It would have no chance of self-reflection, no basis for self-consciousness, no hope for self-disclosure. Internality signifies that we want to look inwardly, that we do not wish to be pulled from ourselves. But if we do not desire inwardness, this lack of desire, too, is how we wish to express our selfhood. A self may be genuine and self-aware, yet another may be its opposite. One may be authentic, whereas another may be a charade. In order for us to be true to ourselves (the Socratic *know thyself*), we must be involved with our own becoming, we must be enmeshed and ensnared with the depth of our own existence uncontaminated by external forces. Although aware of the world, we must look within in order to be attuned with ourselves. Thus, the presence or absence of internality is the key by which we judge the truth of ourselves, because internality means that never are we less alone than when we are by ourselves.

Apart from the benefits our age confers, technology has not made us wise. It has given us a greater awareness of the world at the expense of our own self-awareness. It has helped us to understand so much of the world, and yet so little of ourselves. Information about the physical world has increased immensely through the aid of technology, but not so the knowledge of the self. In fact, self-knowledge is not possible unless we are conscious of the need for it, and the emergence of this consciousness is obstructed by various needs and desires. Disenchanted and inwardly divided, the self disintegrates in the presence of technology because the latter multiplies the devices of self-evasion, leaving us with little more than an empty shell, a hard, crusty exterior and no interior, like an insect without its entrails.

The limitations imposed upon all of us are augmented by the rise of collectivism, by the loss of the individual to gregariousness, and above all by the predominance of the mass over the individual. Technology as it functions in the modern world renders the individual superfluous because it facilitates the negation of our image in our own eyes. Therefore, a world increasingly given over to a technology that rivals human contact is a world in which we have very little significance. Individuality, self-awareness, freedom--all of which may be equated somewhat as the same thing--are ignored, covered over or denied. What results is the lack of spontaneity and the lack of initiative for the self, without feeling, innate power or impulse.

Although technology is utilized as one means to perfection, it does not and cannot concern itself with the highest of all perfections--the perception of the self. Since technology cannot achieve this form of perfection, it is of minor importance. Whatever advantages we receive from technology are secondary to being, since the perfection of being is our primary concern. If technology distracts us from this task, then it proceeds at our expense; we may attain mastery over all things through technology but nevertheless lose the sense of being itself, because we are duped into thinking that we may solve our problems of existence simply by mastering things, by merely controlling everything external to ourselves. Therefore, failing to look inwardly causes a twofold shortfall in our knowledge, for not only do we lose the knowledge of ourselves, but also, by so doing, we lose the knowledge to know others.

Needless to say, technology challenges man's development. Because it is presumed to supply all of our answers, we cease to ask questions of anything other than technology. Hence, we have no need to reflect about ourselves or the world to any appreciable degree because in a technological age everything is taken for granted. Whatever technology can do is accepted as a given. Wherever we might wish to go is already made available. And when everything is done for us, we will do nothing ourselves. It is for this reason that psychoanalysis flourishes in our time. Because we lack a backbone, we have come to possess an almost intrinsic ambiguity. We prefer that we not act for ourselves but that others act for us. Hence, we have forgotten or wish to forget that if we are to be free, we must manifest it through activity.

Because we allow ourselves to be overwhelmed by technology, we lose the sensitivity to perceive and understand great problems. We dilute our strength, yet though our strength is dissipated on the one hand, we are ignorant of the possibility of sublimation on the other. And in our shortsightedness, we project our potentialities upon technology, which confers upon us in turn the destructive fate of self-objectification--we imagine ourselves to be mere physical objects. But what is worse is that we make a mechanical object our paradigm, as when we say that the computer thinks or when we describe it as a psychological machine. The dehumanizing effects of technology reduce us to a function of itself because technology impoverishes us. It isolates us from the real world, making us incapable of dealing with our own experience. Technology does not concern itself with subjects, but with objects; it is not concerned with the being of human beings. Thus, technology encumbers being; it objectifies us. We witness the alienation of the self because technology has made us superfluous. No matter how serious the threat of alienation or inauthenticity may be, a more serious threat faces us, the threat of degradation, the process by which we are reduced to a thing. As life becomes increasingly uninteresting, we become increasingly insignificant.

The massiveness of a technological age is alienating because technology imparts a feeling of impotence. Even with sophisticated weaponry, our participation has been reduced to pushing a button. But alienation also increases the desire to find satisfaction only within one's imagination, to live only within an imaginary world; thus technology has the tendency to make us schizophrenic. It encourages us to escape from reality and to be at odds with ourselves. Because we are free, we may be anyone in fantasy and no one in reality. But we must be conscious of alienation, if we are to do anything about it. We must suffer through it, if we hope to better ourselves despite it. If technology does not alienate us, it will surely overwhelm us, and if we gave in completely to it, we would gladly abandon our liberties to a system which gives us a sense of meaning. Since devotion demands sacrifice, demagoguery will always have a future.

Perhaps technology is no more than a poor compensation for our empty and weakened state, a condition which is aggravated when we, creatures disadvantaged by nature, take pride in our achievements, as if we seek solace in our own prowess. In addition to all its imperfections, we can say emphatically that technology is a vehicle of self-deception, an agent of delusion, the means through which life's great questions of life and death, morality and justice, beauty and truth are ignored and falsified. These great queries, strengthened by defiance and betrayal, are left behind unconsidered in the wake of a passing technology. Because technology encourages us to strive after what we are not while denying to us who we are, we are left all the more poorer by its presence, since the deeper we are immersed in technology, the more we are removed from ourselves.

It goes without saying as a criticism of our time that we need to make exceptional people, who are not completely satisfied with themselves or the world. We need people who will qualify and not quantify morality, who will create values in conformity with life by defining value as discipline, for nothing worthwhile is ever achieved through permissiveness. While not ignoring the rights which have been conferred, we must define ourselves more through the obligations we owe, to ourselves, to our family, to history. No matter what we do, we must strive for excellence. But this also means that we must be willing to change, to question our time-hallowed truths, because we will never create anything new as long as we continue to adapt what already exists; when it is possible to copy, nothing new is ever created.

This is not to say that we can live in a world without technology. Certainly, such a world holds no solution to the problem, for surely if one is nearsighted, glasses would be most helpful. In order to reap advantages from technology without jeopardizing individuals, we must use it to benefit well-being to the extent that being will not be disadvantaged. Although we cannot rid ourselves of technology, we must perfect ourselves in the face of its presence, that is, we must create individuals who transcend the phenomena of our own devising by creating a world that makes this possible. We must internalize both ourselves and our culture. Since rarely does change have great beginnings, we must achieve it through a directive of single-minded purpose by mastering internality. When we create this world and these men, everything else will be superfluous, but such a scheme will be long and dangerous, and will be filled with obstacles and pitfalls. To travel by this route will lead to many hardships, above all, the hardship of self-denial. Although we can easily lose our way, we can take comfort in the signposts of discipline, self-reliance and a sense of belonging, which themselves are helped by an awareness of history, family and collective memories. So let us begin our journey, and as we proceed along the way, hopefully we will recognize that ultimately all we have is who we are because all that we really have in the world is ourselves.

Notes

1. Hume, *A Treatise of Human Nature*, 2nd ed. P. H. Nidditch of the L. A. Selby-Bigge edition (Oxford, 1978), bk. I, pt. 4, § 6 (p. 252).

2. Jaspers, *Man in the Modern Age*, trans. Paul, p. 171. The same idea is expressed by Sartre, *Being and Nothingness*, trans. Barnes, p. 575.

3. Kierkegaard, *The Sickness Unto Death*, trans. Howard V. and Edna H. Hong (Princeton, 1980), pp. 33-34.

Chapter 10

Human Bondage: Technology and a Technological Artifice

When driven by a technological age, technology is the supreme vehicle to dependency. It is the means by which we surrender self-control, the manner by which we seek subjection. Not only do we find solace in the fact that technology is a mechanism of our own making, but it is the very means by which we seek bondage, which is a circumventing way of saying that we no longer wish to be free. And with dependency goes the feeling of irresponsibility, whose presence is offered as proof that indeed freedom does not exist. Although ontologically free, as Rousseau so well points out, we shy from freedom because it is too much of a burden. We bind ourselves to technology because we perceive it as a great enterprise. We enchain ourselves to its compelling force in order that we may share in its awesome power. Without thinking of the consequences, we eagerly carry out the mandates of technology and anxiously await each and every new phase in its development.

To the aware, the technological age comes as no surprise. We have been preparing a very long time for its arrival because its historical roots lie deep in European history. Although classical heritage and Germanic blood lie at the foundation of Western culture, the influence of Christianity has had a far more compelling influence on the development of Western technology. Because the latter received sustenance by being nurtured in Christianity's womb, it profited from Christianity's progressiveness, which was formed and developed through all the Christian centuries from late antiquity to the modern age. Western technology, propelled by a boisterous optimism, has benefited from the belief that humanity has a final and all consuming destiny to fulfill, and like Christianity, technology takes pride in its accomplishments, blindly empowered by the conviction that everything is right with the world so long as one believes. Above all, technology and technological progress have acquired Christianity's notion of inevitability because they are presumed to be unalterable, above all, unavoidable. Because we believe that things will come to pass regardless of what we do, we assume it is hopeless for us to exercise freedom when we wish to do otherwise. This sense of inevitability contains within itself the idea of the futility of resistance, the utter uselessness of saying no to any of technology's manifestations. Like Christianity, technology is depicted as an absolute that would be ludicrous to challenge or question because there is no higher authority.

Technology has succumbed also to the christianization of time, that is, it has adopted Christianity's linear perception. This concept of time as linear, originally devised in medieval monasteries so that monks could work and pray at precise times of the day, later lent its hand to the development of modern technology by insisting that time is not only linear, but also capable of fragmentation into smaller and smaller components, making an accurate reckoning of time possible. And the mechanical clock, an invention of the fourteenth century, further reinforced Christianity's inherent linearity by demonstrating that time could indeed be laid out in a straight line from beginning to end. Time is no longer thought to be cyclic, as nature has it. Because linearity perceives time teleologically, it also means that there can be no retreat, no falling back, to a former state.

Christianity gives the feeling, however groundless, that the world and man are improving, just as Christianity sees itself as indisputable evidence of progress that all men, if they conform to its dogma, will achieve as a mass what no individual could achieve by himself; for Christianity makes overtures for the equality of all men, placing everyone on the same plane because everyone is made in God's image. And so, too, technology has acquired this progressiveness. Biased toward its own success, it never doubts the result of its endeavors. Since it is built on the premise that there will be progress, it follows that we absolutely must be inclined toward its successful conclusion. It tells us that there is no salvation unless we side with technology because beyond its protection is damnation. Whereas Christianity's idea of linearity moves on to a final and decisive end from which there is no appeal, conceptualized in the Second Coming and the Last Judgment, technology's concept never reaches its end, since it is composed solely of means.

Technology also acquired a process of organization from Christianity, characterized by both a systematization and a methodology; that is, Christianity gave technology both a perspective and a method. This systematic methodology imposes a structural unity upon the world, enabling it to bring together everything, even those things intended to be left in a solitary state. Thus, technology becomes a most powerful master. It is not a typical master from whom we desire an increase of self-reliance or discipline or strength, but one we desire to be totally dependent upon. We would rather not act, unless accompanied by technology every step of the way, and we make technology the vehicle of all our thoughts, actions and hopes. Because we complement technology rather than use it to complement ourselves, we have reduced ourselves to being its counterpart.

But the greatest legacy Christianity has given to the development of Western technology comes in the form of its utilitarianism, since Christianity is a pragmatic, not a contemplative religion. Its premise is that we can be saved through conduct, not through reflection. It requires that we act, not that we think. Its goal is that good deeds will save us, even if we do not know why we should perform them. Likewise, technology is utilitarian, pragmatic, active, always on the move, doing all manner of things, forever seeking to make more of itself. Never does technology stop to reflect. Never does it pause to ponder where it came from or where it is going. All it knows is that it must act, even if it ultimately achieves nothing substantive. Whenever opportunity shows its face, technology must seize it. Technology takes precedence over other ideologies and cuts across political and cultural lines with its own ideology. Technology reigns like a queen bee while an army of workers gives it whatever it needs. Technology tolerates all manner of things as long as it incorporates them. It promotes an orthodoxy that all is well so long as it triumphs, but this does not preclude technology from challenging all non- or pre-technological phenomena. Technology wages a war against the infidel, the unbeliever, the other. Even the acceptance of the bare rudiments of technology ultimately challenges what is not technological, and ultimately results in acceptance of everything else technology has to offer. There should be no illusions about how far technology can go. A case in point is electronics, which no matter where it is used, leads to computers, and computers lead to an automated environment. Each new phase that our age makes available is accepted without question because no one person can stop the wheels from turning. Even mechanized agriculture, wherever it is applied, leads to the same restrictive options: chemical fertilizers, pesticides and erosion.

Since morality in the broadest sense of the word is conformity to prevailing practices, which are manifestations of a culture's values, there is nothing that we value more in our time than technology. It helps to define the conditions under which we live, since we are attuned very much to every pulse of its heartbeat. And because technology has become the predominant morality of the modern age, this predominance is one reason why technology is easily transformed into a religion, although it contains within itself all of the basic characteristics of any religion as already enumerated above. Technology purports to rescue us from danger, to give us a feeling of righteousness, to guarantee eternal life, to deliver us from evil. All it requires is obedience to its will, and thus, we have resigned ourselves to fate, to the wishes of the gods; but resignation frequently brings fatalism, which more often leads to treachery than to heroism.

Nevertheless, the old religions, as bad as they were, were kinder. They may have been the means to persecution, alienation and death, of senseless crusades and inquisitions, but they gave a glimmer of hope, at least in the next world if not in this one. Technology, on the other hand, is without kindness because it has no mercy, because the very act of intensification so characteristic of technology denudes the world of love. Unlike God who is merciful even to the worst of sinners--because forgiveness is a divine quality--technology is without mercy simply because it is not a personification of God at all, but a demon dressed in divine clothing. Like the Israelites who worshipped idols, requiring child sacrifice or prostitution, we too prostrate ourselves before technology and gladly sacrifice not only ourselves but our children as well. Ultimately, we care little if technology has no kindness. We are concerned primarily with control, to the exclusion of everything else; and excessive applications of control lead to self-righteous portrayals of domination.

Although the exploitation of nature is as old as man himself, it is impossible for us to say whether, if previous ages could have created an age for themselves similar to our own with all its inexorable ties to technology, they would have done so. But we have done so. Indeed, when it became apparent that such an opportunity could be realized, we did not hesitate in the least to bring all manner of techniques, devices and machines into the world. We filled up the world with everything unnatural. Because we have no difficulty adjusting to a technological age, it is apparent, as we said above, that we have been patiently preparing ourselves for its arrival. And now immersed in it, we are very much accustomed to it. We manifest no hesitation, experience no doubt, nor criticize any premise of it. The present age is the result of a great plan in the grand scheme of things, signifying a stratagem of the utmost proportions.

We are attracted to whatever possibilities lie before us, but we can grasp only at what is present, or at what is perceived to be present. A world without technology offers us little to be excited about. We ignore anything unless it has a technological underpinning. Noticeably, there is an appreciable difference, in fact all the difference in the world, between technology as the means to well-being and a technological age as the means of domination. It is not the former that is to be distrusted, but the latter. It is not the benefits of medicine or agriculture or industry that we should question, but the mentality from which these benefits allegedly spring forth, the premise, believed to be undeniable and incontestable, that every manifestation of life, from conception to death, from the inception of anything to its completion, must have a technological foundation. Although beliefs are often assumed to be facts, we should not be fooled by the idea that technology is what makes life or cultures possible. Because we are extremely tied to technology, since our age is now overwhelmingly technological, it is not surprising that we believe technology to be the giver of life itself. This mutual relationship is our basis for assuming that technology is an ontological phenomenon; but to call technology ontological, as does Heidegger, is in itself a denial of the role of freedom in human action. For if technology is not the result of history and culture--both of which originate because of freedom--then freedom plays no part in its creation, that is, freedom is totally unnecessary to technology and all its functions. Once we have rid ourselves of the burden of freedom, we can assume that all manner of things are assignable to a preconceived destiny.

This confusion between the affirmation of freedom and its denial is characteristic of technology, but it is also characteristic of Western culture, and especially of Christianity. It is evident, for example, in Augustine's idea that we are free, but free only to sin, that is, we are not free not to sin.[1] What Augustine means, of course, is that we are free from sin when we obey God, which brings to mind St. Paul's expression (Romans 6:20-23) that we are most free when we are most a slave. And so, too, is the reasoning of technology, the reasoning that we must always accept whatever is offered. We are presumed free only when we comply with what is given, only when we do what we are told, and thus we are expected to act like children. We are expected always to agree, never to question.

Of all the technologies known to man, modern technology is unique, notably because it makes everything it touches impassioned and intense. It perceives nothing in moderation; it gives the world no tranquillity. Possessing no sense of balance, it never avoids extremes nor honors limits. Always itching to impose, it invades tyrannically everywhere. Although it is true to say that all people, from prehistoric times to the present, have made use of technology to improve life, it is also true that we live in a time when we are more concerned with contrivance than with well-being. We purposely fabricate ways to utilize technology, and we use it to excess, as though we killed an insect with a hammer. We believe we should not do anything without involved and convoluted techniques, without a methodology so inexorably tied to technology that human life itself would be impossible without it. Thus, we question the usefulness of anything unless it is totally embedded in technology. Because only complete absorption concerns us, we care less for the ends to anything than for the means because only the latter can hold our attention; only the very act of engagement itself can sustain us.

The uses to which we put technology indicate that we have confused freedom with bondage, believing that we can retain independence while being tied to a mechanism. Although it is true that we have severed ties from many bonds, we have also anchored ourselves anew to many others. Our dependency is far-reaching. If we have been liberated, for example, from the necessity of becoming great, we have substituted the necessity of becoming great consumers of material goods. If we have been liberated from manipulation by the pulpit, we have submitted instead to manipulation by the media. If we have been liberated from an ignorance of the workings of the world, we have accepted instead an ignorance of ourselves. If we have been liberated from the need for people, we have instead subjugated ourselves to the need for machines. We do not even interact with each other unless technology is the intermediary. We are bored silly if left with just our imaginations. We are more at a loss in a technological age than in former ages because we have rendered ourselves helpless without it.

The desire of an enlightened humanity has always been the ability to build a meaningful future not dominated by ideological conflicts. Despite attempts to achieve this goal, its success is still remote. The great myth of the present age is that technology and its democratic systematization will fulfill this dream. But the basic conditions which characterize us, notably, freedom, openness to being, and the uncertainty that shrouds existence, all of which define us as indefinable, could just as easily lead to failure as to success, no matter how lofty or worthy our goals might be. Above all, we must resist falling into a dull and listless state of being, simply because we adhere to the familiar out of fear of finding something worse; but we must also not be afraid of failure. We may not achieve an absolute good, but we can strive to produce some good. The good we perceive must lead to change, to the ability to profit from struggle and perfection while not leaving essentials behind.

In order to feel secure in a world filled with danger, we customarily rely upon technology to build a wall around ourselves. We seek safety in its protection. But its use creates a world with little hope; it cannot afford us an easy means of escape, a way of leaving the protection of our walled fortress without encountering difficulty. In our present condition, deliberate acts of defiance and their concomitant confrontation rarely happen, except if they conform to technology's manner of doing things, that is, if they adhere to technology's methodology or conform to its democratization. Thus we have embarked on yet another phase of human failing. Since we easily accept whatever technology has to offer, we limit freedom for the sake of progress. Because we readily surrender self-control when confronted by technology, we transform technology, intentionally or accidentally, into a most subtle form of oppression. When we do this, we have little need for change, the absence of which makes us submissive, compliant and obedient. Because change, which is necessary to development and achievement, is an indispensable quality to attaining anything, particularly anything worthwhile, its lack symbolizes an empty listlessness, a dispirited and weakened state, a disorientation. The passing of time itself changes things by a natural process independent of man, but the change we dread is the change needed in order for us to will something to happen, to be made different. And different we dread to become.

The difference we need to create in ourselves is contradictory to our age, even loathsome. Technology has caused softening, dependency, degradation, disengagement and escapism--all of which impart a loss of meaning. We must do the opposite, for we must be brave, self-reliant, resolute, and self-aware, that is, we must become hard. We must not look at technology's values, but through them, questioning every aspect of their manifestation. If they promote well-being, we should keep them. If they do not, we should discard them. And we should decide their usefulness without pity. We should not fear inconvenience, that is, we should not be afraid to struggle, since struggle helps to make us resilient and adds fuel to the fire of the will. Moreover, we must possess determination, which is a supreme quality; we must believe in ourselves in order to achieve anything worthwhile. But we must also cultivate a sense of belonging, a sense of family as well as an awareness of culture, race and humanity. We must know how we fit into the world and its past, and be willing to take command. We must develop qualities of virtue and excellence, such as courage, dedication and integrity, which will help to prepare us for the rigors of life. But despite our preparation, we must also leave the present age behind. We must cross the chasm of indifference into a meaningful future. We must supplant the technological age with a humanistic one, even though humanism will not eliminate suffering or disappointment or death. It will only make us stronger. Hopefully, it will also make us kinder. And when we have crossed this chasm, the present age we leave behind will fade away just as the night dies with the coming of dawn.

Notes

1. Augustine, *On Nature and Grace*, ed. Whitney J. Oates, 2 vols. (New York, 1948), 1, ch. 66 and 79.